PURSUING THE UNITY OF SCIENCE: IDEOLOGY AND SCIENTIFIC PRACTICE FROM THE GREAT WAR TO THE COLD WAR

From 1918 to the late 1940s, a host of influential scientists and intellectuals in Europe and North America were engaged in a number of far-reaching unity of science projects. In this period of deep social and political divisions, scientists collaborated to unify sciences across disciplinary boundaries and to set up the international scientific community as a model for global political co-operation. They strove to align scientific and social objectives through rational planning and to promote unified science as the driving force of human civilisation and progress.

This volume explores the unity of science movement, providing a synthetic view of its pursuits and placing it in its historical context as a scientific and political force. Through a coherent set of original case studies looking at the significance of various projects and strategies of unification, the book highlights the great variety of manifestations of this endeavour. These range from unifying nuclear physics to the evolutionary synthesis, and from the democratization of scientific planning to the utopianism of H.G. Wells's world state. At the same time, the collection brings out the substantive links between these different pursuits, especially in the form of interconnected networks of unification and the alignment of objectives among them. Notably, it shows that opposition to fascism, using the instrument of unified science, became the most urgent common goal in the 1930s and 1940s.

In addressing these issues, the book makes visible important historical developments, showing how scientists participated in, and actively helped to create, an interwar ideology of unification, and bringing to light the cultural and political significance of this enterprise.

Harmke Kamminga was a historian of science and medicine, most recently affiliated with the Wellcome Unit for the History of Medicine at the Department of History and Philosophy of Science, University of Cambridge, UK. She was the author of numerous publications on the history of biochemistry, laboratory medicine and origin of life theories. Her books include *The Science and Culture of Nutrition, 1840–1940*, edited with Andrew Cunningham, *Molecularizing Biology and Medicine* and *Representations of the Double Helix*, both edited with Soraya de Chadarevian. Harmke passed away in 2013.

Geert Somsen teaches history of science and technology at Maastricht University and was Marie Curie fellow at Columbia University from 2014 to 2016. His research focuses on science and international relations. Recent publications include *Neutrality in Twentieth-Century Europe: Intersections of Science, Culture and Politics after the First World War*, edited with Rebecka Lettevall and Sven Widmalm. He is working on a new book, tentatively titled *'Science and World Order': Uses of Science in Plans for International Government, 1899–1950.*

Science, Technology and Culture, 1700–1945

Series Editors

Robert M. Brain
The University of British Columbia, Canada
and

Ernst Hamm
York University, Canada

Science, Technology and Culture, 1700–1945 focuses on the social, cultural, industrial and economic contexts of science and technology from the 'scientific revolution' up to the Second World War. Publishing lively, original, innovative research across a broad spectrum of subjects and genres by an international list of authors, the series has a global compass that concerns the development of modern science in all regions of the world. Subjects may range from close studies of particular sciences and problems to cultural and social histories of science, technology and biomedicine; accounts of scientific travel and exploration; transnational histories of scientific and technological change; monographs examining instruments, their makers and users; the material and visual cultures of science; contextual studies of institutions and of individual scientists, engineers and popularizers of science; and well-edited volumes of essays on themes in the field.

Also in the series

Entrepreneurial Ventures in Chemistry: The Muspratts of Liverpool, 1793–1934
Edited by Peter Reed

Barcelona: An Urban History of Science and Modernity, 1888–1929
Edited by Oliver Hochadel and Agustí Nieto-Galan

Science Policies and Twentieth-Century Dictatorships Spain, Italy and Argentina
Edited by Amparo Gómez, Antonio Fco. Canales and Brian Balmer

Pursuing the Unity of Science

Ideology and Scientific Practice from the Great War to the Cold War

edited by

Harmke Kamminga
(University of Cambridge)

and

Geert Somsen
(Maastricht University)

Routledge
Taylor & Francis Group
LONDON AND NEW YORK

First published 2016
by Routledge

2 Park Square, Milton Park, Abingdon, Oxfordshire OX14 4RN
711 Third Avenue, New York, NY 10017

Routledge is an imprint of the Taylor & Francis Group, an informa business

First issued in paperback 2017

Library of Congress Cataloging in Publication Data
Names: Somsen, Geert. | Kamminga, Harmke.
Title: Pursuing the unity of science : ideology and scientific practice from the Great War to the Cold War / [edited] by Geert Somsen and Harmke Kamminga.
Description: Farnham, Surrey ; Burlington, VT : Ashgate, 2016. | Series: Science, technology and culture, 1700–1945 | Includes bibliographical references and index.
Identifiers: LCCN 2015042126 (print) | LCCN 2015045389 (ebook) | ISBN 9780754640356 (hardcover : alk. paper)
Subjects: LCSH: Science—Philosophy. | Science and the humanities. | Interdisciplinary approach to knowledge. | Science—Europe—History. | Science—United States—History. | Logical positivism.
Classification: LCC Q175.3 .P867 2016 (print) | LCC Q175.3 (ebook) | DDC 509/.041—dc23
LC record available at http://lccn.loc.gov/2015042126

ISBN: 978-0-7546-4035-6 (hbk)
ISBN: 978-0-8153-6680-5 (pbk)

Typeset in Garamond Premier Pro
by Swales & Willis Ltd, Exeter, Devon, UK

Contents

Figures

Contributors

Dr Tim Boon (PhD University of London) is Head of Research and Public History for the Science Museum Group in Britain and a historian of the public culture of science. He is responsible for developing the Museum's research programme. His exhibitions include *Health Matters* (1994), *Making the Modern World* (2000) and *Oramics to Electronica* (2011). His first book, *Films of Fact*, was published in 2008 (London), followed in 2013 by the edited collection *Material Culture and Electronic Sound* (Washington DC, with Frode Weium). He has published extensively on science in film and television, on museums and museum practice, and on technology and music.

Peter Galison (PhD Harvard University) is Joseph Pellegrino University Professor in physics and history of science at Harvard University. His books include *How Experiments End* (Chicago, 1987), *Image and Logic* (Chicago, 1997), *Einstein's Clocks, Poincaré's Maps* (New York, 2003) and *Objectivity* (New York, with L. Daston, 2007). His films are *Ultimate Weapon: The H-Bomb Dilemma* (with Pam Hogan, 2000), *Secrecy* (New York, with Robb Moss, 2008, Sundance) and *Containment* (2015, Full Frame) – about guarding radioactive materials for the 10,000 year future. Galison collaborated with William Kentridge, on a multiscreen installation, *The Refusal of Time* (2012). He is finishing the book "Building Crashing Thinking" about the self and modern technologies.

David A. Hollinger (PhD UC Berkeley) is Preston Hotchkis Professor of History Emeritus at the University of California, Berkeley. He is an elected Fellow of the American Academy of Arts and Sciences, is a former President of the Organization of American Historians, and has served as Harmsworth Professor of American History at Oxford. His six books include *After Cloven Tongues of Fire* (Princeton, 2013).

Rafael Huertas (MD, PhD Universidad Complutense de Madrid) is a research professor at Spanish National Research Council (CSIC). In recent years, he has been working on the history of psychiatry and public health in Spain. He has been head of the Department of History of Science (History Institute, CSIC) and President of the Spanish Society for the History of Medicine.

Jeff Hughes (PhD Cambridge University) is a senior lecturer at the Centre for the History of Science, Technology and Medicine, University of Manchester. He has published widely on the history of radioactivity and nuclear physics and is the author of *The Manhattan Project: Big Science and the Atom Bomb* (Duxford, 2002, and New York, 2003). His current research projects include the history of the Royal Society in the twentieth century and the development of post-war air navigation systems.

Harmke Kamminga (PhD University of London) was a historian of science and medicine, most recently affiliated with the Wellcome Unit for the History of Medicine at the Department of History and Philosophy of Science, University of Cambridge, UK. She was the author of numerous publications on the history of biochemistry, laboratory medicine and origin of life theories. Her books include *The Science and Culture of Nutrition, 1840–1940*, edited with Andrew Cunningham (Amsterdam, 1995), *Molecularizing Biology and Medicine* (Amsterdam, 1998) and *Representations of the Double Helix* (Cambridge, 1995), both edited with Soraya de Chadarevian. Harmke passed away in 2013.

George Reisch (PhD University of Chicago) is a historian of the philosophy of science. He published *How the Cold War Transformed Philosophy of Science* (Cambridge, 2005) and is currently writing a book on Thomas Kuhn and James Conant. He is managing editor of *The Monist*.

Fernando Salmón (MD, PhD Universitat Autònoma de Barcelona) is a professor of the history of medicine at the University of Cantabria (Spain). His research interests include the history of the modern hospital in the twentieth century and the history of medicine in the Middle Ages. His latest book (in collaboration with Michael R. McVaugh) is volume 14 of the *Arnaldi de Villanova Opera Medica Omnia* (Barcelona, 2014).

Vassiliki Betty Smocovitis (PhD Cornell University) is a professor of the history of science in the Departments of Biology and History at the University of Florida, Gainesville, Florida. Her research has focused on the history of evolutionary biology and the unifying properties of the discipline. She is the author of *Unifying Biology: The Evolutionary Synthesis and Evolutionary Biology* (Princeton, 1996).

Geert Somsen (PhD Utrecht University) teaches history of science at Maastricht University and is a Marie Curie fellow at Columbia University for 2014 to 2016. His research focuses on public representations of science and on science and international relations. Recent publications include *Neutrality in Twentieth-Century Europe: Intersections of Science, Culture and*

Politics after the First World War, edited with Rebecka Lettevall and Sven Widmalm (New York and London, 2012). He is now working on a book, tentatively titled 'Science and World Order: Uses of Science in Plans for International Government, 1899–1950'.

Bert Theunissen is a professor of the history of science and the director of the Descartes Centre for the History and Philosophy of the Sciences and the Humanities at Utrecht University. His research focuses on the history of the life sciences in the nineteenth and twentieth century. His current project centres on the history of livestock breeding in de Netherlands in the twentieth century.

Rein de Wilde (PhD Maastricht University) is a professor of science and politics and former Dean of the Faculty of Arts and Social Sciences, Maastricht University (2005–2015). He has published on the disciplinary history of sociology, technological utopias and the art of future-telling.

Preface

This volume has had a long and turbulent prehistory. Its first conception came from Harmke Kamminga, my co-editor, who had become interested in discussions and actions related to the 'unity of science' in the interwar period. What particularly attracted her to the subject was the combination of epistemology, left-wing politics and hands-on scientific practice that galvanised the unity of science pursuits of the time. Harmke's enthusiasm was contagious, and together we organised a workshop, found a publisher and started to work toward an edited volume. We got well underway, but before it was over, the process slowly grinded to a halt.

There were various reasons for this, but one was, I believe, that the subject was so close to Harmke's heart. It brought together the greatest intellectual passions that she had developed during her lifetime: the history of the life sciences, the philosophy of science and a commitment to radical politics. Harmke had always been a prolific writer and editor, but this combination of engagements was too large to fit into a single chapter and yet too important to simply give up. As a result, she struggled with her own contribution and the book-making process petered out rather than coming to a definite end.

The volume had virtually gone into hibernation when it was suddenly shaken up by Harmke's sad and untimely passing away in 2013. It seemed to me, however, that this should not be the end but rather the impulse to revive and finish the work. To my joy and gratitude all of the contributors and the publisher agreed, and we jointly brought the existing chapters up to date and the volume to completion. The result is now before you. Harmke's chapter is still missing, but the volume as a whole is very much hers. Harmke did not believe in an afterlife, but I am glad that her work lives on this way.

Geert Somsen

Acknowledgements

The workshop from which most of the following chapters emerged was made possible by generous support from the Huizinga Institute for Cultural History. Editing of the volume was supported by a Marie Curie International Outgoing Fellowship within the 7th European Community Framework Programme.

Chapter 1
Introduction

Harmke Kamminga and Geert Somsen

'Are the sciences one or many?', Thomas Kuhn asked in an essay in 1976.[1] Kuhn himself did not answer the question definitively, but since then scholars have increasingly stressed the second of his options. Myriad studies have highlighted the diversity of scientific practices and their contingency on specific circumstances. A whole body of work has revealed the local, the particular and the culturally variable contexts that constitute science in its various branches.[2] Recently a number of philosophers have begun to try to turn the tide and seek new common ground beneath the differences.[3] But whether they will succeed

---

1 Thomas S. Kuhn, 'Mathematical vs. Experimental Traditions in the Development of Physical Science', *Journal of Interdisciplinary History* 7 (1976), 1–21.

2 Localist and contextualist historiography is by now too large to summarise. For overviews see David Livingstone, *Putting Science in its Place: Geographies of Scientific Knowledge* (Chicago, 2003) and Jan Golinski, *Making Natural Knowledge: Constructivism and the History of Science* (Cambridge, 1998). For systematic discussions of issues of place and disunity see Peter Galison, 'Introduction: The Context of Disunity', in Peter Galison and David J. Stump (eds), *The Disunity of Science: Boundaries, Contexts, and Power* (Stanford, 1996), 1–33; and Adi Ophir and Steven Shapin, 'The Place of Knowledge: A Methodological Survey', in Adi Ophir, Steven Shapin and Simon Schaffer (eds), *The Place of Knowledge: The Spatial Setting and Its Relation to the Production of Knowledge*, special issue of *Science in Context* 4 (1991), 3–21. The disunifying effect of this historiography has been accompanied by expressions of concern about loss of coherence, for instance in James Secord (ed.), *The Big Picture*, special issue of *British Journal for the History of Science* 26 (1993), 387–483; and Robert E. Kohler et al., 'Focus: The Generalist Vision in the History of Science', *Isis* 96 (2005), 224–51.

3 See Jan Faye, *Rethinking Science: A Philosophical Introduction to the Unity of Science* (Aldershot, 2002); Shahid Rahman, John Symons, Dov M. Gabbay and Jean Paul van Bendegem (eds), *Logic, Epistemology, and the Unity of Science* (Dordrecht etc., 2004); and John Symons, Olga Pombo and Juan Manuel Torres (eds), *Otto Neurath and the Unity of Science* (Dordrecht, 2011). In the 1990s philosophers of science had also moved away from the assumption of a fundamental unity of science. See John Dupré, *The Disorder of Things: Metaphysical Foundations of the Disunity of Science* (Cambridge, MA, 1993) and Alexander Rosenberg, *Instrumental Biology or the Disunity of Science* (Chicago, 1994) presented normative arguments for the disunity of science.

or not, it is unlikely that the notions of universal scientific method and the fundamental sameness of all things scientific will return in an unqualified manner. In this volume, we do not deplore that state of affairs or reject the insights that have been gained. On the contrary, we suggest that they raise a major historical issue: if some kind of *disunity* now seems the normal state of science, why did scientists and philosophers in the past hold so strongly to its unity? More specifically, why did 'unity of science' become such a potent ideal and widely used slogan in the interwar period?

The explicit promotion of the unity of science in the interwar years is best known from the Unity of Science Movement set up by the philosophers of the Vienna Circle. Otto Neurath, Rudolph Carnap and their fellows strove to unify all science into a single body of knowledge, stripped of any 'dogmatic', 'metaphysical' and otherwise unscientific contaminations. Their ideas and activities have been studied extensively, and we now know a great deal about the goals, contexts and political dimensions of these philosophical pursuits of unification.[4] This kind of attention has not been paid, however, to a host of related, simultaneous projects that took place within the natural sciences. For at the same time that Neurath et al. developed their epistemological ventures, physicists launched reductionist programmes linking the understanding of all phenomena to the behaviour of subatomic particles, biologists advanced the evolutionary synthesis in an effort to unify disciplines inside and beyond the life sciences, and scientists from several backgrounds campaigned for unified science as a basis for planning and social reform.[5] Unity of science

---

[4] Nancy Cartwright, Jordi Cat, Lola Fleck and Thomas Uebel, *Otto Neurath: Philosophy Between Science and Politics* (Cambridge, 1996); Jordi Cat, Nancy Cartwright and Hasok Chang, 'Otto Neurath: Unification as the Way to Socialism', in J. Mittelstrass (ed.), *Einheit der Wissenschaften* (Berlin, 1991); George Reisch, 'Planning Science: Otto Neurath and the *International Encyclopedia of Unified Science*', *British Journal for the History of Science* 27 (1994), 153–75; Symons, Pombo and Torres, *Neurath and the Unity of Science* (n. 3); A.W. Richardson, *Carnap's Construction of the World: The Aufbau and the Emergence of Logical Empiricism* (Cambridge, 1998); Michael Friedman and Richard Creath (eds), *The Cambridge Companion to Carnap* (Cambridge, 2007); Thomas Uebel, *Empiricism at the Crossroads: The Vienna Circle's Protocol-Sentence Debate* (Chicago, 2007); Friedrich Stadler, *The Vienna Circle: Studies in the Origins, Development, and Influence of Logical Empiricism* (Vienna, 2001). The latter and other volumes on the manifold activities of the Vienna Circle are issued regularly under the auspices of the Institute Vienna Circle in Vienna.

[5] On particle physics and reductionism see Chapter 4, this volume; and J. Hughes, *The Manhattan Project: Big Science and the Atom Bomb* (Cambridge, 2003); P. Galison, *Image and Logic: A Material Culture of Microphysics* (Chicago, 1997); Daniel J. Kevles, *The Physicists: The History of a Scientific Community in Modern America* (New York, 2001 [1977]); E.F. Keller, 'Physics and the Emergence of Molecular Biology: A History of Cognitive and Political Synergy', *Journal of the History of Biology* 23 (1990), 389–409.

pursuits were widespread and manifold and by no means the monopoly of the Viennese philosophers.

As with the Vienna Circle, whose project extended beyond logical positivist philosophy to include larger social and political goals, there are many indications that the undertakings by scientists in the same period embraced much more than the unification of knowledge. The principles that they thought unified science – whether a form of knowledge, a common method or a general attitude – were deemed to have wider applications as well, which would exert their own unifying effects. Scientific approaches, for example, were viewed as potential instruments for government that could cut through political divisions. Scientific values were taken to define modern culture and connect the past, present and future of human civilisation. And the unity of the scientific community was projected as a model for international relations, sounding a rallying cry against the divisive forces of nationalism and war. Specific instances of these developments have received historical attention, but there has been no systematic scrutiny of the diverse efforts of scientific unification.[6] This collection makes a start with that.

On the evolutionary synthesis see Chapter 3, this volume; and Vassiliki Betty Smocovitis, *Unifying Biology: The Evolutionary Synthesis and Evolutionary Biology* (Princeton, 1995); Ernst Mayr and William B. Provine (eds), *The Evolutionary Synthesis: Perspectives on the Unification of Biology* (Cambridge, MA, 1980). On science and social reform see Chapters 6, 7 and 8, this volume; and Gary Werskey, *The Visible College: A Collective Biography of British Scientists and Socialists of the 1930s* (London, 1988 [1978]); William McGucken, *Scientists, Society, and State: The Social Relations of Science Movement in Great Britain 1931–1945* (Columbus, 1984); Patrick Petitjean (ed.), *Politically Engaged Scientists, 1920–1950: Science, Politics, Philosophy and History*, special issue of *Minerva* 46 (2008), 175–270.

[6] On the belief in impartial scientific government see e.g. William E. Akin, *Technocracy and the American Dream: The Technocrat Movement, 1900–1941* (Berkeley, 1977); Tony Judt, *Postwar: A History of Europe Since 1945* (New York, 2005) 67–77; Robert N. Proctor, *Value-Free Science? Purity and Power in Modern Knowledge* (Cambridge, MA, 1991). On modernism and science see e.g. Bruce Clarke and Linda Dalrymple Henderson (eds), *From Energy to Information: Representation in Science and Technology, Art, and Literature* (Stanford, 2002). On science as a model for internationalism see e.g. Waqar Zaidi, 'Technology and the Reconstruction of International Relations: Liberal Internationalist Proposals for the Internationalisation of Aviation and the International Control of Atomic Energy in Britain, USA and France, 1920–1950' (PhD diss., Imperial College, London, 2008) and Rebecka Lettevall, Geert Somsen and Sven Widmalm (eds), *Neutrality in Twentieth-Century Europe: Intersections of Science, Culture, and Politics after the First World War* (New York, 2012). Excellent, and rare, studies of cultural meanings of science are Andrew Jewett, *Science, Democracy, and the American University* (Cambridge, 2012) and David A. Hollinger, *Science, Jews, and Secular Culture: Studies in Mid-Twentieth-Century American Intellectual History* (Princeton, 1996).

Unification in context

The world in which interwar pursuits of the unity of science emerged was one of deep division and splintering antagonism. Fragmentation not only seemed to characterise the academic realm of ongoing specialisation, but also, all too obviously, the world of international relations and social strife. Amidst the First World War, Bolshevik Revolution, nationalist agitation and the threat of another global conflict, the claim that any part of human endeavour was fundamentally united was bound to have broader import. It was precisely because science seemed to transcend these many oppositions that it carried a message to them. If the world was divided, science could unite it, and pursuing scientific unity inevitably had political and cultural as well as epistemological implications.

To be sure, this had also been true for other times – for unity of science was not a singularity of the interwar period, as Peter Galison makes clear in Chapter 2. The famous 'school of 1847' scientists around Hermann Helmholtz and Emil du Bois-Reymond, for example, strove for a unification of physiology with Newtonian physics at the same time that they tried to establish a unified German state. Knowledge and liberal nationalism were here closely intertwined. Similarly, in mid-twentieth century America, different views of unity in physics coincided with different visions of democracy: a pyramidal reductionism being associated with the rule of law, a quilt-like map of science with the protection of diversity. Interwar unity of science pursuits, however, were distinguished by their enormous scope and their often overt political commitment. Unifying science was strongly (but not exclusively) associated with left-wing orientations; it was a distinctly international undertaking, often with internationalist intentions; and it was deemed to be meaningful for an extraordinarily wide range of human endeavours.

Both the broad scope and the intensity of these pursuits were a response to the perception of profound political and cultural crisis that was rife after 1918. The promotion of unified science as a rational basis for social planning and international cooperation formed a collective answer to the feeling of general devastation that followed the Great War. It thus presented a counterweight to the fast-spreading cultural pessimism that foresaw the downfall of Western civilisation rather than new opportunities for progress. As such, unity pursuits were part of a greater modernist project, both as regards their rationalism and their forward-looking mentality. Indeed, associations with functionalist architecture and abstract painting were explicitly made.[7] Important strands of modernism regarded science as possessing uniquely progressive, unifying powers, and scientists simultaneously drew upon and contributed to these perceptions.

---

7 C.H. Waddington, *The Scientific Attitude* (Harmondsworth, 1941); and Peter Galison, 'Aufbau/Bauhaus: Logical Positivism and Architectural Modernism', *Critical Inquiry* 16 (1990), 709–52.

The pursuit of the unity of science acquired its greatest momentum with the economic crisis of the Great Depression and the growth of fascism in the 1930s. Belief in the general potential of science joined scientists from a broad political spectrum against the 'irrationality' of the free market and the purported regressiveness and barbarism of the Nazis. Scientists' subsequent engagement in the war effort, on a scale unprecedented in material and ideological terms, also fuelled optimism about the role of science in post-war reconstruction.[8] Ideals of the scientific organisation of a united world informed the establishment of new international agencies such as the United Nations Educational, Scientific and Cultural Organization (UNESCO), the Food and Agricultural Organization (FAO) and the World Health Organization (WHO).[9] Global social unity through science seemed on the verge of being realised, and for many scientists these were years of unparalleled hope. Some even hailed the atomic bomb for the unifying effects it would have, as it left countries no choice but to share the new knowledge and collectively manage its applications.[10]

Yet soon after the Second World War the ideology of unity of science began to be marginalised. Strong and explicit unions between science and politics were increasingly associated with totalitarian abuse, the Soviet Union becoming the primary example with the onset of the Cold War. By this time, many natural scientists turned away from left-wing activism, seeking instead to strengthen the autonomy of science and keep political interference at bay. As mentioned earlier, pursuits of scientific unity did continue in areas such as American physics, but their ideological connotations were both different and more hidden than during the 1930s. Similarly in the philosophy of science, the entrenchment of Cold War divisions narrowed the Unity of Science Movement down to an internalist epistemological programme, stripped from its former political dimensions. The movement had come to the US with its ideology intact, but was reduced to 'pure' logical empiricism in the McCarthy era, as George Reisch shows in our final chapter. The Cold War terminated the most ambitious unity pursuits.[11]

---

8 By the end of the Second World War the belief in scientific planning was shared left to right. Tony Judt has called it 'the political religion of postwar Europe' – see Judt, *Postwar* (n. 6), 67–77.

9 G. Archibald, 'How the S Came to Be in UNESCO', in P. Petitjean et al. (eds), *Sixty Years of Science at UNESCO* (Paris, 2006), 29–34. Aant Elzinga, 'UNESCO and the Politics of Scientific Imperialism', in Aant Elzinga and C. Landstrom (eds), *UNESCO and the Politics of Scientific Imperialism* (London, 1996), 89–131.

10 Paul S. Boyer, *By the Bomb's Early Light: American Thought and Culture at the Dawn of the Atomic Age* (New York, 1985), chs. 2–3.

11 For comparable developments in the history of science, see Anna-K. Mayer, 'Setting Up a Discipline: Conflicting Agendas of the Cambridge History of Science Committee, 1936–1950', *Studies in the History and Philosophy of Science* 31 (4) (2000), 665–89; and 'Setting Up a Discipline, II: British History of Science and "the End of Ideology", 1931–1948', *Studies in the History and Philosophy of Science* 35 (1) (2004), 41–73.

Unifying issues

What happened in the interim – between the Great War and the Cold War – and how specific pursuits of the unity of science developed during this time constitutes the subject matter of this book. A number of questions guide the authors' analyses. How was the unity of science translated into specific programmes of research as well as activities outside the laboratory? What functions did appeals to unification serve in the scientific, social, philosophical and political realm? How was the unity of science ideology situated culturally? Such issues are pertinent to particular unification programmes and to the unity of science movement as a whole. It will become apparent that there were many interconnections between the different unification endeavours, stemming from personal contact, shared circumstances and sometimes coordinated action. Links with the Unity of Science Movement of the Vienna Circle will appear frequently as well, even though there is no indication that the other endeavours derived top-down from its epistemological initiatives. On the contrary, the book shows that the Viennese programme was but one part of a much larger enterprise from which it was singled out for scholarly attention only later – especially after the pursuit of the unity of science had been narrowed down to a purely philosophical project.

The actors discussed here were a mixed bag of people. Some of them were working scientists, others science writers, popularisers or journalists. Some were documentary or propaganda makers and still others had history or philosophy as their main occupation. Just like the members of the Vienna Circle, most of whom were not card-carrying philosophers but natural and social scientists who took to – and redefined – philosophical issues, the actors appearing in this volume often moved between different fields. In an era where history and philosophy of science were not yet firmly institutionalised and demarcated and, for instance, science journalism was hardly a recognised profession, such movements were easy and natural. What was common to these various roles was that our actors served as public spokespeople for science. They represented what science could achieve and how it should be allowed to do that while speaking to wider audiences and debating in a greater public sphere.

Unity of science pursuits had important ideological dimensions, but it should not be suggested that all action was driven by deeply held convictions or lofty idealism. On the contrary, personal, institutional, disciplinary and even national interests played a pervasive role, as the subsequent chapters show. At the same time, we do want to draw attention to the fact that, in the period discussed, people did commonly see their own and others' actions in terms of grand ideological schemes. Ideology gave meaning to action, and the political significance of particular choices was often explicitly expressed. These dimensions have not yet been given the attention they deserve. Studies of 'science and ideology' have commonly focused on situations outside Western democracies where 'totalitarian' doctrines left their imprint on scientific practice – Aryan physics and Lysenko's biology being the paradigmatic examples. Describing these situations has usually implied their condemnation. By contrast, this book seeks to highlight ideological dimensions of science precisely *in* Western, democratic societies, without the implication that they corrupted 'proper practice'.[12] The systematic distinction between ideology-infected totalitarian science and non-ideological free science is itself, we suggest, an ideological construct, largely of the Cold War.

Diversity in unity

The volume opens with a long-term overview of pursuits of scientific unification that immediately illustrates their ideological dimensions. Moving from the early nineteenth to the late twentieth century, Peter Galison describes three of the most famous and ambitious endeavours in the unity of science: the programme of Hermann Helmholtz, Emil du Bois-Reymond and the German 'school of 1847', the Vienna Circle's Unity of Science Movement (the only one literally going under that name) and the various pursuits of unification of and through physics taking place in Cold War America.

Each of these projects sought the unity of knowledge on a different epistemological basis and, accordingly, carried different political overtones. The Germans wanted to unite the life sciences and physical sciences by subjecting them to the same laws of nature and, similarly, strove for political unification of the various German states under a single constitution – both nature and politics should be one under the law. For the Viennese philosophers, it was not law but language that would unite science. Once stripped of all metaphysics and expressed in concrete, everyday terms, knowledge could be connected across the disciplines and assembled in the *International Encyclopedia of Unified Science*. Its non-hierarchical organisation mirrored the international and social order that the Unity of Science Movement sought to embody and advance. Mid-twentieth century American physicists like Victor Weisskopf held much more reductionist views of unification. All phenomena rested on physics, just as all politics was based on constitutional principles. But contemporary colleagues, like Philip Anderson, disagreed. Their view of a quilt of collaborating subdisciplines saw strength in diversity and celebrated the *e pluribus unum* that was also thought to empower America. In each of these cases, Galison shows, conceptions of political unification did not derive from visions of scientific unity or vice versa. Rather they formed two inseparable dimensions of the same pursuit.

---

[12] For an earlier attempt to stretch these limits, see Mark Walker (ed.), *Science and Ideology: A Comparative History* (London, 2003).

The book's diachronous opening is followed by nine synchronous case studies, all situated roughly between Great War and Cold War. All address several dimensions of scientific unification pursuits – epistemological, political, cultural – but they differ as to which of these dimensions serve as the starting point. The chapters are grouped accordingly.

Chapters 3 and 4 begin by considering unifying practices and discourses within specific disciplines and move on to examining the wider dimensions of these endeavours. Betty Smocovitis takes on the famous 'evolutionary synthesis' that was launched in interwar biology. Following the career of one of its main architects, Julian Huxley, she analyses its various levels of unification. The synthesis was meant, first of all, to unify hitherto loosely connected subdisciplines, such as systematics, ecology and genetics, under the theoretical umbrella of evolution – a unification that was institutionalised in new biology journals and organisations. This operation also put the field in the position of a 'central science', connecting even non-biological fields, like cosmology and astrophysics, which started to adopt evolutionary perspectives. Finally, the synthesis turned evolution into a general world-view, most conspicuously in the form of Huxley's 'evolutionary humanism', which subtly steered between left-wing materialism and right-wing vitalism and mysticism. It was this centrist political manoeuvring, Smocovitis argues, that allowed this unification ideology to survive the Cold War – and to continue in the work of such late twentieth-century scientists as Carl Sagan and E.O. Wilson.

In chapter 4 Jeff Hughes looks at another great interwar unification pursuit, namely that of atomic physics, a field regularly advertised as laying the foundations for all other sciences. This, at least, is how Ernest Rutherford presented the reductionist programme of nuclear research that he and his group carried out at the Cavendish laboratory in Cambridge, UK. Atomic physics would not only unify the sciences, it would also unite researchers all over the globe – and particularly in the British Empire – because of the common ground found in the results of careful experimentation. These lofty claims, however, concealed a much messier research practice, where it was hard to stabilise measurements and where researchers in different places repeatedly challenged the outcomes of each other's experiments. Rutherford could only contain the controversy by appeals to collegial solidarity and requests to keep the differences private. What unity was eventually achieved came less from profound agreement on principles or methods than from the standardisation of commercially available instruments. Rutherford subsequently moved his rhetoric to more mundanely economic and national themes, but the claim of fundamental unification continued to be trumpeted by science writers and journalists.

The next two chapters approach unity of science endeavours from the angle of their internationalist intentions. Geert Somsen discusses a series of initiatives by Dutch scientists who promoted themselves as guardians of international

David Hollinger looks at the reception of Otto Neurath's Unity of Science Movement in the United States, analysing the well-prepared grounds on which it fell. In fact, the goal of a more scientific society centred on unified knowledge had been pursued long before Neurath presented it to American intellectuals on a visit in 1936. John Dewey, who had been central to this effort, could even reuse an article of 1910 as his contribution to Neurath's *Encyclopedia of Unified Science*. The American movement was older and broader (including much more social science and Marxism as legitimate knowledge) but also more specific in its general cultural outlook. The idea of a scientific attitude, and the universalism associated with it, was pitched against a narrowly Anglo-Saxon and protestant conception of American society and in favour of an ethnically diverse America. A science-based society meant making space for Jewish and central European immigrants, as Hollinger shows in Sinclair Lewis's and Thorstein Veblen's celebrations of the Jewish scientist. Even if the specific epistemological project of Neurath and his American colleagues would eventually fail, this larger reconceptualisation proved to be successful, as testified by the strong presence, by mid-twentieth century, of Jewish intellectuals in the American academic community.

George Reisch looks at the Unity of Science Movement in the United States as well, but follows how it developed after its arrival. Like Hollinger, he points out that it was initially warmly welcomed because the Viennese goals coincided with American ones, also politically. Most East Coast intellectuals were left-leaning and shared the socialist, scientistic orientation of Neurath and his fellows. By the 1950s, however, the project had been narrowed to a strictly logical-philosophical undertaking, stripped of its political dimensions. This change did not happen automatically, Reisch argues, but was the result of deliberate McCarthyite actions. Left leaning became suspect, and J. Edgar Hoover's FBI investigated leaders like Carnap and Philipp Frank and interrogated people around them. It was soon understood that political engagement could seriously damage one's academic career, or worse. And if this was not generally clear, the message was brought home personally by philosophers such as Sidney Hook, who had turned fervently anti-communist. Both in public and private correspondence, Hook insisted that anything smacking of socialism was a danger to society and should be dealt with accordingly. It was in this 'climate of fear' that the Unity of Science Movement lost its broader appeal and eventually withered away.

Chapter 2
Meanings of scientific unity
The law, the orchestra, the pyramid, the quilt and the ring

Peter Galison

Contrary to its almost constant invocation in the Sunday sermonising of science and philosophy of science, unity is not an invariable part of science.[1] There is no single strand of unity talk, no 'unity idea' passed down across the ages from Thales to the modern syntheses of biology or physics. Instead, the unity of science – as a regulative part of scientific theorising – is a relatively modern phenomenon, one that came to widespread celebration to a very large extent in Germany around the middle of the nineteenth century. This modern unification enthusiasm drew its strength from the idealist philosophers, but gained real force as a scientific, metaphysical and political programme around the German 'Professors' Revolution' of 1848. Since that time, to be sure, unity has come to mean different things at different times at different places – there is a *dis*-unity to the genealogy of unity. But in writing a history of our scientific present, our starting point lies more in Bismarck's Berlin than in Anaximander's Athens.

On the occasion of the fascinating collection of essays assembled by the editors of this volume, it is perhaps worth stepping back to sketch out the larger contours of these diverse unities. Such an overview might help situate some of the specific movements for the unity of science; more proximately it might shed some light on our current situation, in which science seems to many scientists to be more quilt than pyramid. And it may be that the image of a ring will prove of more use than either. But this is truly a sketch, a view more from 30,000 feet than through the jeweller's loupe of case studies.

The law

The liberal nationalist German scientists of the nineteenth century cut a large swath through organised knowledge, not least those young medical scientists

---

1 Portions of this chapter are drawn from Peter Galison, 'Introduction: The context of disunity', in Peter Galison and David J. Stump (eds), *The Disunity of Science: Boundaries, Contexts, and Power* (Stanford, 1992), 1–33; idem, 'Constructing Modernism: The Cultural Location of *Aufbau*', in Ronald N. Giere and Alan W. Richardson (eds), *Origins of Logical Empiricism* (Minneapolis, 1996), 17–44; and idem, 'The Americanization of unity', *Daedalus* 127 (1998), 45–71.

who, around the attempted revolutions of 1848, strove simultaneously for a new physiology unified with physics and for a unified German nation. Among the key figures were Emil Du Bois-Reymond, Hermann Helmholtz and Ernst Brücke, who – in close association with Karl Ludwig in Marburg – were instrumental in forming the 'school of 1847', which sought to ground physiology in Newtonian physics. The three had much in common: Prussian-born and middle-class, similar in age, they came through the University of Berlin together and had all studied under Johannes Müller, the Professor of anatomy and physiology.

At the time of the 1848 uprisings, fought under the banner of 'Unity and Freedom', these physiologists were politically active liberals supporting the calls for a unified German nation and a constitutional assembly. When even the moderate reforms gained in 1848 were reversed already in 1849 and Prussian hegemony was consolidated, they mostly turned away from overt politics and focussed on building their scientific careers. They did maintain their ideal of a unified nation as espoused by the liberal nationalists, who aimed for a constitutional monarchy through moderate political reform of existing bureaucracies. The unification eventually achieved under Bismarck, in the form of a nation ruled centrally from Berlin, suited their purposes as members of the professional middle-class establishment.

Emil Du Bois-Reymond was born of a French-Swiss immigrant father and the daughter of a minister serving the French colony in Berlin. He studied philosophy, theology, psychology, logic, metaphysics, anthropology, botany, geography and meteorology; to boot he was a student of Müller. After Müller's death in 1858, his Chair was split into two and Du Bois-Reymond was elected to the Chair of Physiology. Among Du Bois's allies was Karl Ludwig, the foremost practitioner of physiology in Germany, training a generation of the leading figures in the field at his institute in Leipzig. Hermann Helmholtz, too, counted among Du Bois's friends and he, too, famously ranged widely and broadly – from fundamental studies on thermodynamics, including his crucial work on the conservation of energy, to the physiology of the senses. He invented the ophthalmoscope, contributed importantly to electrodynamics, reconsidered the foundations of geometry and grappled seriously with the philosophical meaning of science. The empirically oriented physiologist, Ernst Brücke, who was driven by politics to Zürich after 1848, led a generation to the study of physical physiology.[2] Rudolf Virchow, who went on to create 'cellular

---

2 The literature on the relevant parts of late nineteenth-century German science is vast. The following works contain extensive bibliographies and I refer the reader to them: David Cahan, *Hermann von Helmholtz* (Berkeley, 1993); Klaus Christian Koehnke, *The Rise of Neo-Kantianism*, trans. R.J. Hollingdale (Cambridge, 1991); Timothy Lenoir, *The Strategy of Life* (Chicago, 1989); idem, *Instituting Science: The Cultural Production of Scientific Disciplines* (Stanford, 1997); Kathryn M. Olesko, *Physics as a Calling* (Ithaca, 1988). Particularly relevant are also: Sven Dierig, *Wissenschaft in der Maschinenstadt: Emil Du Bois-Reymond und seine Laboratorien in Berlin* (Chicago, 2008); Laura Otis, *Müller's*

pathology' and strove to unite medicine and society through scientific endeavour, was also associated with them and sympathetic on many broader issues. Politically more radical than the 1847 group, Virchow was forced to resign his post and leave Berlin after 1848. He joined the democratic *Volkspartei* which, unlike the liberals, aimed for the overthrow of existing political structures and for the establishment of a German republic.

Du Bois-Reymond was to become in many ways the epitome of the German academic establishment, and he worked hard, over a lifetime, to join up the historical, political and moral functions of science. At the University of Berlin he was rector in 1869–70 and in 1882–83. He was a member of the Prussian Academy of Sciences and became one of its permanent secretaries in 1876; in this capacity he had a great deal of political influence. As his career flourished, Du Bois increasingly promoted the development of pure science for the sake of a greater German culture. He also articulated the fundamental questions that science could and could not expect to solve. His famous 1872 *Ignorabimus* speech, insisting that some questions are unknowable, was hailed and derided on presentation and remembered for its slogan 'we cannot know', proclaiming the limits of science.[3] Some took Du Bois's placement of science to put up defensible borders within which science could be practised freely; others argued, on the contrary, that such limits left space to religious knowledge.

Du Bois's 1872 speech took the aim of all knowledge to be its grounding in the 'mechanics of atoms' – it was a belief that he and his circle held deeply. Newtonian *law* lay at the foundation of all science. According to Du Bois, what was needed was the explanation of sensory impressions in terms of central forces alone. And this would require a dramatic reform of medicine, pathology and physiology, unifying the whole under the banner of the physical sciences. Du Bois asked: what is the ultimate nature of matter? We cannot know. That is the lower limit of science. What is consciousness? We cannot know that either – that is the upper limit of science. Science, therefore, existed as a

---

Lab (Oxford, 2001); David Cahan, *Hermann von Helmholtz and the Foundations of Nineteenth-Century Science* (Berkeley, 1993); Richard Kremer, *The Thermodynamics of Life and Experimental Physiology, 1770–1880* (New York, 1990); and Henning Schmidgen, *Die Helmholtz-Kurven. Auf der Spur der verlorenen Zeit* (Berlin, 2010). On contemporary projects to collectively unify knowledge, see Peter Galison and Lorraine Daston, 'Scientific Coordination as Ethos and Epistemology', in Helmar Schramm, Ludger Schwarte and Jan Lazardzig (eds), *Instruments in Art and Science. On the Architectonics of Cultural Boundaries in the 17th Century* (Berlin, 2008), 296–333; and Peter Galison, *Einstein's Clocks, Poincaré's Maps* (London, 2003).

[3] What I say about this draws much from the remarkable but unfortunately still unpublished work of Keith Anderton, 'The Limits of Science: A Social, Political, and Moral Agenda for Epistemology in Nineteenth-Century Germany' (PhD diss., Harvard University, 1993).

unified, safe region between these two seas of the unknowable. Such a safe middle landscape was central to liberalism as it was understood by the stalwarts of 1848. These bourgeois revolutionaries aimed for freedom of conscience, expression and questioning; a freeing up of the economy, industry and education. They sought a system of belief and of knowledge building that was, by its very nature, pitted against feudalism, absolutism and religious orthodoxy. German liberals wanted a unified *Rechtstaat* in politics, a constitution of fixed laws and they wanted a fixed grounding of laws for science. Among the German nationalist liberals, unity oscillated easily and often from politics to scientific culture to philosophy.

As a vehicle for their unity aspirations, the liberal nationalists saw none better than the *Gesellschaft Deutscher Naturforscher und Ärzte* (Convention of German Scientists and Doctors). It was a perfect symbol of scientific and political drawing together, shifting, like a movable feast, among the German states. For years after 1848, the language of science and cultural unity self-consciously replaced that of political unity in public, at least until the 1860s.

The rule of law in science stood for constitutionalism for the nation – and the liberals understood their science to signify a particularly German-inflected form of free society. They argued that science instantiated freedom, but freedom under law, not the lawless, 'mechanical' (as opposed to the German 'organic') form of French freedom with its echoes of revolutionary anarchy. Benjamin Theodor Thierfelder opened the Convention of German Scientists and Doctors in Rostock in 1872 with the rallying cry: 'The victory of German weapons is a victory of the German spirit'.[4] According to Anderton, it was widely held among the liberal scientists that the Convention had been of key importance in keeping the ideal of unity alive in German hearts for decades. Helmholtz, ending the Innsbruck Convention of 1869 on the verge of German unification, took the aim of science to be the demonstration that all of nature would be unified around the basic laws of mechanics. Politics was not far away: the 'cardinal virtues of German science' could provide 'a means of deepening the sense of unity throughout our country'. In the same year, Du Bois concluded, 'The German unity which is finally nearing completion was first conceived in the German universities'.[5] Just after the rise of the German nation, Virchow looked back in a Convention address entitled 'On the Tasks of the Natural Sciences in the New National Life of Germany', where he contended that 'the

---

4 Anderton, 'Limits of Science' (n. 3), 201, who draws it from Heinrich Schipperges, *Weltbild und Wissenschaft: Eroeffnungsreden zu den Naturforschersversammlungen 1822–1972*, Schriftenreihe zur Geschichte der Versammlungen deutscher Naturforscher und Aerzte, vol. 3 (Hildesheim, 1976), 46–7.

5 Anderton, 'Limits of Science' (n. 3), 203–4. Quoted from H. von Helmholtz, 'Aim and Progress', in *Popular Lectures on Scientific Subjects*, trans. E. Atkinson (New York, 1873), 397, and from E. Du Bois-Reymond, 'Ueber Universitaetseinrichtungen', in *Reden*, vol. 1 (Leipzig, 1912), 368.

present glory of the German nation would have been impossible if, in loyal, untiring work, the German universities had not been at their posts, if within the circle of the universities the national idea had not been fostered again and again, until it went out to all the world'.[6]

It should be noted that physicists – even when they believed that Newton's laws would serve as the unifying basis for all science – had in mind different pictures of what mechanics was. Some chose to emphasise energy over forces (as did Helmholtz himself) and others fastened on most to the least action principle (as did late Helmholtz and Max Planck). But the most remarkable shift took place with the spectacular reversal articulated by Wilhelm Wien in 1900. Wien began arguing at that moment that the unification of all things scientific was to be predicated not in terms of mechanics, but even mechanics itself was to be accounted for in terms of the underlying theory of electricity and magnetism. But these unification strategies cannot be read in isolation from the changing world around them. In the unification of science, the construction of the vast unified mechanical *Weltanschauungen* paints a picture in which all is gears, pulleys, heat, engines – an imprinted vision of the mechanical, productive force of mid-nineteenth-century Germany and Britain. It was then displaced in 1900 by the electrodynamic world picture: a world in which all was electrons and their fields. Here we ought to see not the gears and steam engines of Victorian England, but Prussian dynamo halls and power transmission lines. We should see lamps torn from the walls so electric lights could take their place; carriage roads ripped up end to end to make way for electric trams. This was the street scene correlate of the electromagnetic world picture.[7]

But whether unified science was based in energy, in least action or in central forces, whether it began with charged electrons or inert mass, the great aspiration of the late nineteenth century, in the view from Berlin, was for a science that was unified under the law. The Prussians' goal was to understand the nature of that law, grappling with what we could know a priori and inserting what we could not know in advance by scrupulous experimental inquiry into everything from the outside world to our innermost senses. It was a task that all at once was of central importance to philosophy, to science and to the politics of a law-bound freedom for a united Germany.

The orchestra

Our overflight continues. In a second period, roughly from 1918 to 1939, the most widespread talk of unity of science came from another quarter altogether.

---

[6] Cited in Anderton, 'Limits of Science' (n. 3), 205. Quoted from R. Virchow, 'Aufgaben der Naturwissenschaft', in Karl Sudhoff (ed.), *Rudof Virchow und die Deutschen Naturforscherversammlungen* (Leipzig, 1922), 108–9.

[7] See Dierig, *Wissenschaft*, (n. 2).

Based in Vienna, not Berlin, unity stood for something quite different from the drawing together of previously separated German states, the elimination of tariffs or the establishment of a nation. Instead, unity of science resonated with a kind of high modernist construction, a resolutely *international* form of political ideal, one predicated on a shared language of scientific reasoning. Scientists hailed their own unification across domains as the very instantiation of rationality: beyond nation, race, or partisan political squabbling. In the hands of the largely (but not exclusively) socialist-leaning philosophers of science and their scientist allies, unification of science formed a bulwark against fascism, nationalism, clericalism and mysticism.

Undoubtedly the new and restless science-inflected philosophy was important for early twentieth-century unity talk. But intellectual – even political – rebellion was dwarfed by the awfulness of the First World War. The mechanised destruction of millions of lives put talk of rationality, science and unity under an intense spotlight. One of the responses to slaughter by machine-gun and gas was a drive to create a new society, or, more specifically, to harness the mechano-rational to a new, more rational, more scientific way of living. One indicator of this self-consciously modernist desire was a burst of post-war German-language journals with '*Aufbau*' in the title. The word is of course familiar from construction sites on every street corner of the German-speaking world. (In the late twentieth century there was even a shampoo called 'Aufbau' to convey its building-up power for lacklustre hair.) But back in the years following the European war, *Aufbau* stood neither for everyday construction nor for the frivolity of cosmetics.

Aufbau conjured up a building up anew. Not *re*-construction, but making anew. Here was a conjoint effort in knowledge, in city planning and in economy. Among the dozens of journals and tracts bearing that name, one (*Der Aufbau: Flugblätter an Jugend*) began in 1921: 'What is *Aufbau*? Is *Aufbau* "changing the world"? Evidently not! For what changes the world more than revolution? And yet we say: a revolution creates only the conditions for an *Aufbau*. Change is attached to the given, to the substantial; it is a redirection, a shifting of forces; it is the technical'.[8] Another (*Umsturz und Aufbau*) stated in 1919:

> We have lived through the most unheard of catastrophe that a people has ever encountered. In need and death, in blood and tears there has been only one asset . . . the spirit. To make the spirit fruitful for the renewal of our people is the goal of our series *Umsturz und Aufbau*. . . . We want to remove the rubble of the European killing fields and with heart and head to prepare for the erection of a new humanity.[9]

8 H. Schüller, 'Revolution-Aufbau', *Der Aufbau; Flugblätter an Jugend* 1 (1921), 2–7: 5.

9 *Umsturz und Aufbau* 3 (1919; reprinted in 1974), 22–3.

Yet a third (*Der Aufbau*) – this one co-edited by some of the leading architects and city planners of Red Vienna alongside the polymath sociologist Otto Neurath – insisted in 1926: 'We want to work together on the cultural, social and economic *Aufbau* of society, which is striving from the present unclarity, disorder and chaotic confusion toward clarity and order'.[10] Neurath moved easily back and forth between politics, aesthetics and philosophy of science. At the root of all three, he was sure, lay a rational, rationalised life, engineering efficiency and the coordination of the different special sciences. Neurath likened scientific unification not so much to obedient subjects of a divine (or Newtonian) law, but rather to the harmonious combined playing of the various instruments of an orchestra.[11]

Neurath was one of the founders of what became known as the Vienna Circle – which to many philosophers stands at the beginning of the whole of modern philosophy of science. Along with physicists-turned-philosophers Philipp Frank and Rudolf Carnap, and led by the charismatic Moritz Schlick (the only card-carrying philosopher among them), this small group and a handful of others struggled to strip speech to its logical and empirical foundations and to build knowledge anew from clear, well-founded first principles. This building-up would have a rigorous structural integrity that could be shared beyond the particulars of one person or another – an *Aufbau*.

In the early Vienna Circle meetings, much like Berlin in the 1840s, boosters of unity of science often held back from talking explicitly about governments. Peter Hempel, a participant in some of the logical positivists' assemblies, once told me that there was a rather explicit agreement to 'leave politics at the door'. But if politics stood outside, its terrifying presence hovered like a spectre over the table, only occasionally calling to the assembled.

Neurath, with Carnap and Charles Morris, edited the *Encyclopedia of Unified Science*, where the eminent American philosopher John Dewey contributed a rousing defence of the unity of method, a 'scientific attitude' *(wissenschaftliche Weltauffassung* to the Vienna Circle) that brought disparate specialties together. But Dewey wanted more than scientific results alone: '[T]here is also a human, a cultural meaning of the unity of science'. Beyond the reformation of one's own individual stance towards the scientific method, one's 'efforts are

---

10 'Unser Wollen', *Der Aufbau. Österreichische Monatshefte für Siedlung und Städtebau* 1 (1926), 1.

11 Here and in the following discussion I draw on Peter Galison, 'Aufbau/Bauhaus: Logical Positivism and Architectural Modernism', *Critical Inquiry* 16 (1990), 709–52; and idem, 'Constructing Modernism' (n. 1). On Neurath and political unification, see Nancy Cartwright, Jordi Cat, Lola Fleck and Thomas E. Uebel, *Otto Neurath: Philosophy Between Science and Politics* (Cambridge, 1996). See also John Symons, Olga Pombo and Juan Manuel Torres (eds), *Otto Neurath and the Unity of Science* (Dordrecht, 2011).

hampered, often times defeated, by obstructions due not merely to ignorance but to active opposition to the scientific attitude on the part of those influenced by prejudice, dogma, class interest, external authority, nationalistic and racial sentiment and other powerful agencies'.[12] There were evils afoot, real evils that could be fought only through the adoption of a scientific attitude promulgated through education.

It was relatively easy to agree that unification was good. What unification *meant* was entirely less unanimous. For Rudolf Carnap, the engineering physicist-turned-philosopher, unity of science carried with it a sense rather different from that of Dewey. For Carnap, to demonstrate unity was to exhibit what he called the common 'reduction basis' of different branches of science, such as biology and physics. By this he never meant that biological laws could be replaced by physical laws. Carnap considered that an entirely open scientific question. Nor did Carnap consider the reduction project to be an ontological one in which the entities of biology, say, would turn out to be nothing but the entities of physics suitably arranged. Instead, he argued that *both* biological and physical laws could be expressed in terms of everyday physical terms and procedures. There was therefore, for Carnap, a reduction of *language* quite distinct from a reduction of *laws* or *ontology*. Linguistic reduction was central.

Despite their occasional clashes, Neurath's 'unification' was not so terribly far from Carnap's. Finding a linguistic unity involved the deployment of a common language of everyday terms, the adoption of a system of universally recognisable icons and most importantly an 'encyclopaedic' assemblage of scientific subjects without forcing them into a suprascientific philosophical 'system' such as Kantianism. For Victor Lenzen, the reduction was more explicitly nomological:

> In the face of apparent disunity, developments in contemporary physics inspire the hope that quantum mechanics and the theory of relativity may be united in a single theory. And because of the basic function of generalised physics and the ever increasing development and adaptation of the techniques of specialised physics, the progress of physics augurs well for the unity of all empirical science.[13]

Insofar as quantum electrodynamics represented the prototypical unification then, at least for Lenzen, unification meant precisely the creation of integrated laws.[14] As in science, so in society: coordination through language,

---

[12] John Dewey, 'Unity of science as a social problem', in R. Carnap, O. Neurath and C.W. Morris (eds), *Encyclopedia of Unified Science*, vol. 1 (Chicago, 1955 [1938]), 29–38: 32–3.

[13] Victor Lenzen, 'Procedures of Empirical Science', in Carnap, Neurath and Morris, *Encyclopedia of Unified Science* (n. 12), 338.

[14] On the meaning of unification for Lenzen, Carnap and Neurath, see Galison, 'Introduction: The Context of Disunity' in Galison and Stump, *Disunity of Science* (n. 1).

method, laws – these were the ways of holding the whole together against disintegration.

But as with their predecessors Helmholtz, Virchow, Ludwig and Du Bois-Reymond, there was no way of entirely avoiding political power, classically conceived, when we turn to post–World War I Vienna, where unified science took the form so well known in philosophy of science. Surely no one lived the question of unity with the urgency of the sociologist-philosopher-cultural critic Otto Neurath. For Neurath's Austro-Marxism, it was planning, not nation building and ideology, that was important. He wanted a rationalised and coordinated distribution of goods, farming, production and currency. Soon after the foundation of the Bavarian Republic by the socialist leader Kurt Eisner in November 1918, Neurath came to Munich to urge his thoughts about central planning on Eisner and on the Munich Workers' and Soldiers' Council. After the violent suppression of the Bavarian Republic by the *Freikorps* in May 1919, Neurath was arrested, tried and sentenced, only to be rescued before he landed behind bars – his saviour was an old friend and ally, Otto Bauer, the Austrian Minister of Foreign Affairs. In this political history, as we know from Cartwright, Cat, Chang, Fleck and Uebel, we find an important wellspring for Neurath's outlook on unified knowledge. The virtues of centralisation and unification seemed as apposite in matters of state as in matters of science. But more subtly, Cat, Cartwright and Chang point out that Neurath's political sense of unification never involved a homogenisation or radical hierarchy. Quite the contrary, his vision in the Bavarian Socialist Republic was always of a *coordinated* economy, left substantially in private hands and synchronised rather than brutally assembled into a vertically integrated whole. In science, Neurath's vision of 'orchestration' similarly avoided the coercive implications of science structured from one 'master law' all the way down through the nitty-gritty of applications. Neurath never believed that science would amount to a small set of microphysical equations from which the laws of ecology, economics and psychology might be derived.[15] (Though, as I will point out in a moment, that *was* a goal for one strand of post–World War II nuclear and particle physics.)

I want to return to the fundamental point about this period leading up to the Second World War: unification was a form of cultural politics; both its supporters *and* its antagonists understood the link between scientific unification and international cooperation. So too in other domains of culture: there was a similar fierce attachment to systematicity, unification and removal of ambiguity among modernist architects. Hannes Meyer of the Bauhaus produced a League of Nations plan that would leave no back corridors for the diplomats' secret

---

15 Jordi Cat, Nancy Cartwright and Hasok Chang, 'Otto Neurath: Unification as the Way to Socialism', in J. Mittelstrass (ed.), *Einheit der Wissenschaften* (Berlin, 1991), 91–110; and Cartwright et al., *Otto Neurath* (n. 11).

deals that many believed had precipitated the Great War. Schlick, in 1928, persuaded Carnap to rename his magnum opus, dropping *Konstitutiontheorie* and taking on *Logische Aufbau der Welt* (logical construction of the world): an attempt to unify all knowledge into a vast structure embracing everything from auto-psychological experience to physics, chemistry and sociology. Carnap, as a good neo-Kantian, remained 'neutral' as to the proper starting point of this construction: his *Aufbau* (his construction) could start with an individual's experience; it could start with the collective. It was the systematic building up that mattered; the placing of all knowledge in a uniquely specified place, or absolute commitment to the launching point of knowledge, was far less significant.[16]

Keeping in mind the deep embedding of unity of science within culture and politics, listen to Carnap's own preface to his *Logische Aufbau der Welt*:

> We do not deceive ourselves about the fact that movements in metaphysical philosophy and religion which are critical of such [a scientific] orientation have again become very influential of late. Whence then our confidence that our call for clarity, for a science that is free from metaphysics, will be heard? It stems from the knowledge or, to put it somewhat more carefully, from the belief that these opposing powers belong to the past. We feel that there is an inner kinship between the attitude on which our philosophical work is founded and the intellectual attitude which presently manifests itself in entirely different walks of life; we feel this orientation in artistic movements, especially in architecture . . . of personal and collective life, of education and of external organisation in general. We feel all around us the same basic orientation, the same style of thinking and doing. Our work is carried on by the faith that this attitude will win the future.[17]

If cultural and political allies of the philosophers of science recognised this common form of life, so too was it a weapon-at-hand against the Nazi threat on every level: as a political organisation, as a model for rational decision making, as a unifying cultural apparatus across national boundaries.

Unity was at once political in its anti-fascist pronouncements and in the practices that made up everyday scientific life. This was certainly how the situation looked to the scientists and philosophers drawn to the *movement*: Unity of Science. Theirs was an international vision, a bulwark, its advocates hoped, for linking scientific fields, for forging a common scientific method, for coordinating international organisations and for building a shared scientific language. And perhaps most ambitiously of all, the interwar unifiers aimed to model a form of life around the persona of the rational, technically reasonable

---

16 Michael Friedman, 'Carnap's *Aufbau* Reconsidered', *Noûs* 21 (1987), 521–45.

17 R. Carnap, *The Logical Structure of the World*, trans. R.A. George (Berkeley, 1967 [German edn, Berlin, 1928]), preface.

scientist – all against the powerful and rising enemy of racism, fascism, religious fanaticism, irrationalism and nationalism.

The pyramid and the quilt

Our final view down from our high-altitude flight is of the third period, the violent half century one might call the Long War (1939–89) in which unity of science took on a more imperial and centred quality. Nuclear science arose in strength with the discovery in the 1930s first of the neutron and then of nuclear fission. During and after the Second World War, Los Alamos accelerated this science with prestige, facilities and intellectual labour that only a centrepiece of a world war could provide. From humble, semi-empirical beginnings in radiochemistry, atomic physics and shoestring experiments, nuclear physics morphed again and again – into elementary particle physics with its international community of mobile physicists and city-spanning sites of accelerator laboratories.

During the Cold War particle physics accelerators became sites for symbolic East-West competition. Unity of science came to be a high-prestige search for the building blocks of all things and all sciences. This was by no means undisputed – and the struggle over the right way of understanding the relation of the special sciences to one another had much to say about how scientists saw the present ethos and the future structure of their discipline.

The experience of mass population displacements, mechanised mass death and redrawn political boundaries created the conditions under which *Aufbau*-style modernism took hold as the ideal of unity of science between the wars. To see how differently unity of science looked after the Second World War, recall the Vienna Circle's enemies. The Circle saw itself as fighting the forces of irrationality, forming alliances with modern architects, educators, city planners and at times Austro-Marxist politics.[18] Never politically powerful or even institutionally secure, as time went on their voices were increasingly drowned out by the array of nationalistic forces pitted against them. The drive to a 'Unity of Science Movement' was, for Neurath, Carnap, Philipp Frank and their many allies, part and parcel of a struggle to bring together a rationality and objectivity that would halt racial and nationalistic assaults from dominating the world. Their opponents were Austrian clericalism, entrenched traditional philosophy and later Nazism. Whether through an *Aufbau*, a physicalist thing language, protocol sentences or isotypic images, the Vienna Circle's goal was

[18] Galison, 'Aufbau/Bauhaus' (n. 11); idem, 'Constructing modernism' (n. 1). The Vienna Circle's stance on metaphysics generated a fierce struggle between Carnap and the leading philosopher in interwar Germany, Martin Heidegger – see M. Friedman, 'Overcoming Metaphysics: Carnap and Heidegger', in R.N. Giere and A.W. Richardson, *Origins of Logical Empiricism* (Minneapolis, 1996), 45–79.

to extirpate irrational metaphysics out of the meaningful world. Metaphysics was not some supererogatory flourish, it was the motor behind the worst forms of mysticism and obscurantism that threatened civilisation.

Now I ask you to move your mental image from the unity of science movements in the 1930s to 1947, not in Vienna, but in post-war Cambridge, Massachusetts. Here, the scientific banner flying overhead the newly founded Institute for the Unity of Science is not that of relativity and quantum mechanics, though these might occasionally be invoked by Philipp Frank. Instead, the newly rehoisted scientific flag announced the riveting new, war-boosted interdisciplines: cybernetics, computation, neutronics, operations research, pshycoacoustics, game theory, biophysics, electro-acoustics. The old enemies of interwar Vienna were gone or vanquished: Austrian clericalism and the hollow vestiges of the Habsburg empire did not figure very large in Cambridge, and fascism had been slain – in the scientists' eyes at least – in no small measure because of their intervention. Now these same tools that had won the war promised the world. Cybernetics with its nonlinear feedback was celebrated as offering a way to rewrite the social sciences as well as the natural sciences. The computer's logic was thought to be universal and capable of doing everything from weather forecasting and nuclear weapons design to the resolution of long-standing problems in number theory, to modelling the human mind.[19]

The unification these scientists had in mind was a unification through localised sets of common concepts (a kind of patchwork quilt), not through a global metaphysical reductionism (with the pyramid as model). Were the mathematical and technical features of feedback, control, black boxes, flow diagrams, or extensive forms of a game 'reducible' to nuclear physics? Hardly. Even posing that question about the kinds of problems facing the Institute seems hopelessly inappropriate. With the kind of power these scientists felt they had at war's end, fretting about ontological reductionism must have seemed almost beside the point. As the chemist E. Bright Wilson wrote in 1950 to Gerald Holton, the Institute Secretary: 'The phase of the Institute's work in which I am particularly interested is that which deals with scientific method in its most practical and least philosophical senses'.[20] The Americanisation of unity just after the Second World War was not sited around an isotypic picture language, a physical language, an *Aufbau*, or an orchestration. It was planted around the new sciences of Los Alamos, the Massachusetts Institute of Technology's Radiation Laboratory (the 'Rad Lab'), the stored-program computer of the Institute of Advanced Study in Princeton; this was to be a science unified in

---

[19] See also Peter Galison, *Image and Logic: A Material Culture of Microphysics* (Chicago, 1997).

[20] E. Bright Wilson, Jr to Gerald Holton, 6 October 1950, folder 3, Gerald Holton, Private Papers.

pieces, grounded in common widely applicable concepts, and promising a power beyond dreams.

One last contrast: when the Vienna Circle faced off against theology in their manifesto, they saw mystic obscurantism as a rising threat; however misunderstood or powerless they were, the Vienna Circle aimed to cast millennia of such speculation to the winds. When the Institute for the Unity of Science sent out its first flyers, they made 'Science and Faith' and 'Science and Values' early and long-standing objects of study. In one of the first meetings of the Institute for the Unity of Science, a prominent participant probably spoke for many when he said the public now saw scientists as authorities comparable to the high priests of ancient cults. But the truly staggering feature is not the prominent positive role accorded truth and values, it is that in these first months of the *Pax Americana* this group of scientists, humanists and philosophers could take on God and morality as problems – and fully expect to solve them. There was a brief interregnum after the Second World War before the Cold War began in earnest; this is the period in which the new Institute for the Unity of Science began.

The Cold War brought with it a vastly expanded nuclear community, and the foregrounding of that community, though too long a story to tell here, is coincident with the vision of science as unified through a pyramidical structure in what its own practitioners came to call a *fundamentalist* programme grounded in the great synthesis of quantum mechanics and relativity. This, unlike Neurath's programme, was knowledge built on a base, a base where status, metaphysical primacy and vast scientific resources came to bear.

As Jordi Cat has effectively argued in his fine study of late twentieth-century unification debates, Victor Weisskopf (who participated in nuclear physics and quantum electrodynamics, and directed the European particle physics centre, CERN) was a staunch defender of this pyramidical view. In 1966–67 he put the position in stark terms: 'fundamentalism' was simply the view that 'all sciences are at the end a branch of physics'. 'The fact that everything can be reduced to the Dirac equation is a very important statement. It is the great fact on which science lives and develops. There are certainly more interesting developments in all sciences, but they all use the foundations which are laid by the fundamentalists'.[21]

The raw ambition of statements like Weisskopf's opened a rift. One opponent was Chew, who – as David Kaiser has so beautifully shown – advocated a conjoint effort. On the one side he set the agenda of a long and productive theoretical programme of 'nuclear democracy' that aimed to capture the relations of particle physics without making any entity more fundamental than the others. On the other side he worked hard, through small-group meetings

---

21 Jordi Cat, 'The Physicists' Debates on Unification in Physics at the End of the 20th Century', *Historical Studies in the Physical and Biological Sciences* 28 (1998), 253–99: 262.

and through his macropolitical positions, for a political order opposed to that of McCarthyesque authoritarianism.[22] So it was that even *within* particle physics there was a split that divided a pyramidical notion of unity from that of a more egalitarian 'nuclear democracy'.

But in the long run, it is not from inside particle physics that dissent about pyramid-style unification has been countered most strongly. For my money it is the condensed matter physicist Philip Anderson who best articulated a position against Weisskopfian fundamentalism. Given all that had been accomplished in physics through an understanding of collective phenomena involving many atoms and many electrons, Anderson simply found it an outrage that particle physics had so monopolised status, students and resources. This was not merely symbolic. Anderson was clear that with the attribution of symbolic centrality came quite material consequences: he believed that other domains, especially condensed matter, had suffered at the hands of the imperial grasp of the particle physicists.

Condensed matter physicists were not – in any case not directly – children of Los Alamos. They found their homes away from the massive government-funded laboratories like Brookhaven National Laboratory or Lawrence Berkeley Laboratory, Lawrence Livermore or the new Fermilab accelerator (that Anderson opposed) carved from Illinois farmland. Many, like Anderson, had worked during the war not on the bomb but on electronics projects, radar problems, antenna design. These experiences spoke to tools. But associated with these material practices were the phenomena, and Anderson insisted that the many-bodied world was altogether as compelling – altogether as *fundamental* – as the decay of a meson into two gamma rays or the discovery of a new strange particle. Superconductivity, spin glasses, optical pumping – these were aspects of the physical world that were not, he believed, any less novel, less conceptually novel than the particle physicist's 'elementary' interaction.

In 1967 Anderson replied to Weisskopf and his allies in the Regent's Lecture, republished (in 1972) as 'More is different'. There, he attacked the difference (insisted upon by Weisskopf) between 'intensive' and 'extensive' research: intensive research was supposed to hunt for fundamental laws (read: nuclear physics, particle physics) while extensive research (solid-state physics, plasma physics, most of biology) putatively sought explanations in terms of known fundamental laws. Angry as he was with the particle physicist fundamentalists, Anderson reserved scorn for defeatists from his own field who on the one hand ceded the non-existence of fundamental problems in condensed matter physics and on the other relegated all *other* problems to 'device engineering'. Anderson stated:

---

[22] David Kaiser, 'Nuclear Democracy: Political Engagement, Pedagogical Reform, and Particle Physics in Postwar America', *Isis* 93 (2002), 229–68.

> The reductionist hypothesis does not by any means imply a 'constructionist' one: The ability to reduce everything to simple fundamental laws does not imply the ability to start from those laws and reconstruct the universe. In fact, the more the elementary particle physicists tell us about the nature of the fundamental laws, the less relevance they seem to have to the very real problems of the rest of science, much less to those of society . . . the behavior of large and complex aggregates of elementary particles, it turns out, is not to be understood in terms of a simple extrapolation of the properties of a few particles. Instead, at each level of complexity entirely new properties appear, and the understanding of the new behaviors requires research which I think is as fundamental in its nature as any other.[23]

Analysis was possible, synthesis – to the point of accounting for the properties of biological systems in terms of quantum electrodynamics – impossible. Lower level (more elementary) laws simply did not carry within their reach the properties of the interesting, macroscopic phenomena. In 1971 Anderson campaigned against big particle accelerators. Joining Anderson was James Krumhansl (blamed by Steve Weinberg as being a 'heavyweight' in opposition to the superconducting supercollider [SSC]): 'All men are created equal', he said, and, like the nation, physics should be built 'e pluribus unum'. The unity of physics resides in the 'effort to try to recognise and cross the bridges to colleagues in other areas'. As a nation 'we are still bound largely by the common language of mathematics and by a common approach in our scientific method – measure, analyse, generalise'. Unity 'exists in shared scientific concepts' like the soliton that occurs in many domains of physics. This is the 'nation' of physics.[24] Again – as we have seen so often – there was a swift oscillation between political, metaphysical and scientific reasoning. Unity talk rarely walks alone.

As Cat points out, for Weinberg there were no 'autonomous laws'. Instead Weinberg wanted the 'reductionist attitude [to provide] a useful filter that saves scientists in all fields from wasting their time on ideas that are not worth pursuing'.[25] But there was more to Weinberg's view. He insisted that the reason for hunting fundamental laws should not be sold short. It was *not* that there were great rewards to be had in hunting for the calculational or predictive payoff from quantum chromodynamics – no one was going to calculate in detail how a tomato seed grows from the QCD Lagrangian. Nor should the SSC be built *because* there would be great 'spin-offs' to be had, like Teflon from the atomic bomb project (first put to large-scale application as a sealant for uranium hexafluoride). More precisely, there might be spin-offs, but that was

23 P.W. Anderson, 'More is Different', *Science* 177 (1972), 393–6: 393; reproduced in idem, *A Career in Theoretical Physics* (Singapore, 1994), 1.

24 Cat, 'Physicists' Debates' (n. 21), 279.

25 Ibid., 195 and 299.

beside the point. 'For me', Weinberg wrote, 'reductionism is not a guideline for research programs, but an attitude toward nature itself. It is nothing more or less than the perception that scientific principles are the way they are because of deeper scientific principles . . . and that all these principles can be traced to one simple connected set of laws'.[26]

All this captures very well Weinberg's view: his stance towards the fundamental does not *derive* from the political any more than Chew pulled his conception of bootstrapping from anti-McCarthyism or Helmholtz his unification from liberal nationalism. That said, the very idea category of the 'fundamental' carried with it more than meets the eye. Anderson wanted to rescue the concept from the hands of the particle physicists precisely because the seat explanation is not a throne to be given up lightly. Weinberg knows too that more rests with fundamentality than problem solving alone: 'The reason we give the impression that we think that elementary particle physics is more fundamental than other branches of physics is because it is. I do not know how to defend the amounts being spent on particle physics without being frank about this'.[27] Here is where metaphysics meets the national budget.

Both Long War visions – Anderson-Krumhansl and Weisskopf-Weinberg – carry a political resonance. Both stood for democracy. Weisskopf-Weinberg's pyramid aimed to make science and society proof against intrusion by the lawless and the irrational. Anderson-Krumhansl's linked patchwork aimed to leave room for diversity under a broader rubric of shared methods. Both raised the microcosm of physics into macrocosmic significance. Los Alamos had spawned the pyramid and the Rad Lab the quilt. The struggle for control lasted the whole of the half-century Cold War.

The ring

Since the end of the Cold War in 1989–90, some features of the debate over unity continued unabated, and no doubt they will for a long time to come. But other forces began to act on the relationship of the sciences one to the other, and to the sciences in relation to both industry and the state. On the horizon we begin to glimpse new relationships among the sciences, in which there is neither a lead science (pyramid) nor a few hard-won passages between two at a time (quilt). For the period of the Long War from the early 1940s to the early 1990s, it may well help us to see as well a 50-year struggle between the centralising nuclear spirit of Los Alamos that culminated in two bombs on one side, and the

---

26 Cited in Peter Galison, 'Metaphysics and Texas', review of *Dreams of a Final Theory: The Search for the Fundamental Laws of Nature*, by Steven Weinberg, *The New Republic*, 6 September 1993, 40–43.

27 Ibid.

offspring of many smaller groups that produced hundreds of different techniques and components at the Rad Lab on the other side. But now, on a much broader and deeper scale, scientists have reshuffled their disciplinary cards.

First, the pyramid is crumbling. There are no doubt many reasons for the cancellation of the SSC – there were budget overruns, there were management troubles, there were political miscalculations in the project's Congressional backers, there were disputes within the physics community. But looked at from our perch at 30,000 feet, I think none of these hold a candle to the double supernova that shook the whole of the physicists' contract with the state: the Cold War ended. Gone, in the unexpected flash of a mallet on the Berlin Wall, was the complex set of assumptions that had made political-symbolic sense of the progression from nuclear physics to weapons research to particle physics. Concomitantly, with its comet-like rise, was a worldwide boon in biomedicine, biotechnologies and more proximately the reach of the American National Institutes of Health and, internationally, the rise of multinational pharmaceuticals.

Indeed, the rise of biomedicine and the end of the Cold War hit particle physics at a vulnerable moment – vulnerable in part because of successes that reached an intellectual plateau in the mid-1970s with the establishment of the stubbornly unrevised Standard Model. Hit on all sides, students – and researchers too – began scrabbling for other ways to do physics, other fields to plough. It is instructive to see where they have gone. Some have headed to astrophysics and cosmology which, in part through new systems of land- and space-based instrumentation, has enjoyed some of its most generative years ever, from maps of the microwave background through programmatic searches for gravity waves, exoplanets and gamma-ray bursts. Detectors, modes of analysis, forms of simulation and theories criss-cross back and forth between the very small of particle physics and the very large of the astrophysical. Another site for joint work has been in the nanosciences where surface chemists, engineers, atomic physicists and biologists productively made common cause in hundreds of laboratories across the world. It is not just that they share a particular tool once in a while; no, the collaboration is much deeper. Experiments could not be done without the joint effort to produce certain macromolecules, instruments or systems, or to frame a problem in terms of one or another's discipline's standing intuitions. Even objects are shared: the nanotube to one collaborator is a molecule, to another a wire, to a third a switch.

String theory, in some ways the foster-child descendant of particle physics, no longer carries with it quite the same evangelical fervour of the 1980s. No longer do most practitioners imagine that one day soon someone will publish the one and true fundamental Lagrangian that will truly be a theory of everything. Instead, the field more and more carries its own weight in which mathematicians and physicists are making common cause, illuminating basic relations of algebraic geometry here and quantum field theory there. This is not a matter of passing a new hammer across the disciplinary divide; it is making a new and substantial field together – whether or not it eventuates in the final unification of all science.[28]

In domain after domain, the most generative parts of physics are forming alliances like those of astrophysics/cosmology, nanoscience and string theory. Our earlier sharply hierarchical or flat patchwork metaphors will not do. These are not 'applications' of a single governing law to the more applied domains. Nor are they quilt-like microexchanges with small border regions and stable centres. Instead, it seems to me that we are seeing a topological shift in the intellectual and cultural structure of physics, one that takes a pyramid into a ring: facing outwards everywhere but without a single, acknowledged centre.

My argument throughout this brief excursus has been this: at every stage of the century and a half of modern unity of science movements, science, metaphysics and politics have always been in play at once. It is that confluence that has made unity of science so important for us to understand both in our grasp of the past and our contemplation of the future. Now the physical sciences are in flux, spiralling on one side dramatically towards the marketplace in biophysics, bioinstrumentation, nanotechnology, computer science. By participating so directly in the flow of venture capital, start-up companies and multinationals, one new face of interdisciplinary, mobile and market-driven sciences has found a form of unification. But in the process of achieving such temporal success it is ever harder for science to provide the ideal representation of, or a philosophical argument for, a unified society toward which we might aim. On the other side, the string side, the physical sciences have advanced so dramatically, moving precisely in the opposite direction: by forming an alliance with the most *abstract* of mathematical disciplines. And there, too, for opposite reasons, because of the *dis*-association of string theory from the pragmatic, unification has also pulled away from providing anything like a model for civil society.

More indirectly there is a cultural-political lesson in these current, disorienting shifts. Appropriate to our world, in which the Cold War recedes into the past, we have produced a science that is both deeply and generatively connected – connected in unexpected, untraditional ways – and yet without, and without any prospect of, a single centre.

---

[28] Peter Galison, 'Theory Bound and Unbound: Superstrings and Experiment', in Friedel Weinert (ed.), *Laws of Nature: Essays on the Philosophic, Scientific, and Historical Dimensions* (Berlin, 1995), 369–408; idem, 'Mirror Symmetry: Persons, Values, and Objects', in M. Norton Wise (ed.), *Growing Explanations: Historical Perspectives on Recent Science* (Durham, NC, 2004), 23–66. On the rise of shared techniques in nanoscience, see e.g. Lorraine Daston and Peter Galison, *Objectivity* (Boston, 2007), ch. 7.

Chapter 3

The unifying vision

Julian Huxley, evolutionary humanism and the evolutionary synthesis

Vassiliki Betty Smocovitis

> *By working together, we must lay a conscious basis for a new world order, the next step in our human evolution.*
>
> Julian Huxley[1]

Though it has not seemed as glamorous nor drawn as much scholarly attention as in the physical sciences, the goal of unification has dominated twentieth-century biology, especially in the first half of the century.[2] Comparing their field to the physical sciences, such icons of early twentieth-century biology like John Scott Haldane, William Morton Wheeler, J.H. Woodger and even lesser figures like botanist W.B. Turrill, all wrote extensively in both their scholarly and semi-public and public works of the goal of unifying biology. All attempted to unify what appeared to be a fragmented and immature science and to use biology to ground a coherent world view. Biology, in their vision, could rival the unity of the older physical sciences but also become an autonomous science that could not be reduced to physics. All, however, recognised the difficulty of this project. In 1939, for example, William Morton Wheeler found it so difficult that he remarked that it might take 'a few super-Einsteins to unify biology'.[3]

In the 1930s and 1940s the movement to unify biology accelerated with a group of individuals who attempted to unify the biological sciences within an evolutionary world view. These included figures like Theodosius Dobzhansky, Ernst Mayr, G.G. Simpson, G. Ledyard Stebbins Jr and Julian Huxley. Though they came from different areas of the biological sciences (genetics, systematics,

---

1 As cited in J.P. Sewell, *UNESCO and World Politics: Engaging in International Relations* (Princeton, 1975).

2 V.B. Smocovitis, *Unifying Biology: The Evolutionary Synthesis and Evolutionary Biology* (Princeton, 1996). This chapter develops further some themes introduced in this earlier work.

3 W.M. Wheeler, *Essays in Philosophical Biology* (Cambridge, 1939).

palaeontology and botany), all believed in the unifying ability of evolution and evolutionary theory. Their efforts from the 1930s to 1950s to bring about a synthesis of disparate points of view and representative fields have been variously named by historians of science as the neo-Darwinian synthesis, the modern synthesis, the evolutionary synthesis or simply, and reverently, as 'the synthesis'. They are frequently known as the 'architects' – so designated by Ernst Mayr – of the new synthetic theory of evolution that emerged at this time. The characteristic features of the synthetic theory and its relationship to the historical event of the evolutionary synthesis will not concern us here (discussions can be found elsewhere);[4] instead, I will examine how this attempt at synthesis led to the belief that biology was a unified and central science. At the same time the 'architects' organised, and in fact disciplined, their community through the unifying theory that would serve as the unifying principle of evolutionary biology. The theory formulated in turn, I argue, was consistent with – and would in fact stabilise – a world view that advocated liberal, progressive and secular humanistic beliefs within the interwar political context. This unifying project was carried into the 1950s, the 1960s and well into the latter half of the twentieth century by these biologists' disciples and students like Carl Sagan and E.O. Wilson. In particular, I will explore the interplay of science, world view and ideology (to use historian John C. Greene's popular phrase) by concentrating on the figure best known for his efforts to unify biology within a coherent world view, Julian Huxley.[5]

---

[4] See the historiographic discussion in Smocovitis, *Unifying Biology* (n. 2). So far the most comprehensive work on the synthesis is E. Mayr and W.B. Provine (eds), *The Evolutionary Synthesis: Perspectives on the Unification of Biology* (Cambridge, MA, 1980; 2nd edn, 1999).

[5] See J.C. Greene, *Darwinism Debated: The Adventures of a Darwin Scholar* (Claremont, 1999). See especially his earlier article, reprinted in this volume, J.C. Greene, 'The Interaction of Science and World View in Sir Julian Huxley's Evolutionary Biology', *Journal of the History of Biology* 23 (1990), 39–55. For general background on Julian Huxley, see C.K. Waters and A. Van Helden (eds), *Julian Huxley: Biologist and Statesman of Science* (Houston, 1992); K. Dronamraju, *If I Am to Be Remembered: The Life and Work of Julian Huxley, with Selected Correspondence* (Singapore, 1993); and M. Keynes and G. Ainsworth Harrison (eds), *Evolutionary Studies: A Centenary Celebration of the Life of Julian Huxley, Proceedings of the Twenty-Fourth Annual Symposium of the Eugenics Society, London, 1987* (London, 1989). For special treatment of Huxley's political views, see C. Divall, 'Capitalizing on "Science": Philosophical Ambiguity in Julian Huxley's Politics 1920–1950' (PhD diss., University of Manchester, 1985); R.M. Gascoyne, 'Julian Huxley and Biological Progress', *Journal of the History of Biology* 24 (1991), 433–55; and M. Swetlitz, 'Julian Huxley and the End of Evolution', *Journal of the History of Biology* 28 (1995), 181–217.

Julian Huxley and evolutionary humanism

The famous grandson of Thomas Henry Huxley, also known as 'Darwin's bulldog', Julian Huxley (1887–1975) was the most visible representative of the group of 'super-Einsteins' who would attempt to unify biology. His role in the evolutionary synthesis has been misunderstood by historians of science.[6] He did the most to promote the newly emerging sense of unity in the community and extend this unity to the wider global community at the same time that he offered a coherent framework that preserved an autonomous science of biology. Huxley was a voracious reader and prolific author, an international traveller, an ambitious statesman and an indefatigable promoter of science. Huxley was acutely aware of the fact that biology appeared to be a fragmented, disunified science. He was an active member of the philosophical community, interacted with Bertrand Russell, and was generally well informed of the philosophical status of biology. He was especially sensitive to criticisms of biology since he seemed to inherit the crusading zeal from his grandfather who wished to integrate the sciences in general and biology in particular with traditional humanistic concerns. While the two differed fundamentally on the extent to which one should ground humanistic philosophy (and ethics in particular) in evolution, both had life-long issues with locating 'Man' in the 'Place of Nature' as revealed by recent science and not revelation.[7] If evolution – and biology – were to help inform and reform the life of 'Man', then the science had to be not just unified and coherent, but also progressive.

Julian Huxley's particular version of humanistic theory – a kind of secular belief – thus offered a seemingly contradictory blend of mechanistic materialism within a purposive, progressive and meaningful world view.[8] His world view was first made public in a series of lectures under the title of 'Biology and Man' that he gave shortly after leaving Rice University in 1916. In 1923 he subsequently published the first of many works intended for public consumption that

---

6 See also Smocovitis, *Unifying Biology* (n. 2). Historians of science have generally made little of Huxley's contributions to the evolutionary synthesis, largely because they have sought to understand them in the context of a traditional scientific career. Huxley engaged less and less in formal scientific research as he devoted more of his energies to serving as writer, publicist, and promoter of evolutionary science. His eventual celebrity status did not help give him the credibility or weight with scholars whose contributions were in technical science exclusively.

7 Thomas Henry Huxley wrote a famous essay on this theme with the title *Evidence as to Man's Place in Nature* in 1863, reprinted in T.H. Huxley, *Man's Place in Nature* (Ann Arbor, 1959). See also Julian Huxley's famous articulation of science without revelation in J.S. Huxley, *Religion without Revelation* (London, 1927), revised and reprinted in 1957.

8 See the works in note 5 for more on Huxley and his world view.

promoted this same world view, *Essays of a Biologist*.[9] It was still dominating his world view in his last major collection of essays, *Essays of a Humanist*, published in 1964.[10] For Huxley 'scientific humanism', later more popularly known as 'evolutionary humanism', was more than just an abstraction or a philosophical exercise; it was instead a central feature of his world view, his philosophy of life and simultaneously of his scientific and political endeavours. Over the course of his long career as scientist, popular writer and public servant, in his specific support of endeavours like eugenics and population control, in his moves to reform education and in his support of global movements like internationalism, evolutionary humanism served as a guiding principle. It was in fact a kind of unifying element in his own life.[11]

His progressive visionary world view needed much invention and reworking, however, given the historic events that Huxley had witnessed in his youth. Progress – the hallowed Enlightenment ideal inextricably linked to the historical origins of evolutionary thinking – had been hard to sustain in the modern world.[12] The bloody 'Great War' was accompanied by a growing and widespread sense of degeneration (cultural and biological) and a widespread belief in the decline of the West. A global sense of fragmentation ensued. The need for a more hopeful, optimistic and progressive world view intensified yet further in the 1930s, a decade that saw the Great Depression. Worse, the rise of mass movements like Nazism, fascism and communism, that threatened the autonomy and agency of the individual, made the call for an ethical system that gave a measure of freedom to the individual all the more urgent. For Huxley, a grounding in evolution and the construction of an evolutionary humanism became an imperative for the future of 'modern man'. From the 1930s his growing publication oeuvre was thus fuelled in part by his need to apply evolutionary humanism to cure the ills of the modern world.

By the 1930s Huxley had increasingly turned to such larger themes in evolution, largely giving up his earlier interests in embryology and animal behaviour. Not only was he an active contributor to the literature on evolution, he was also instrumental in helping found a society for experimental biology in 1925. In 1927 he surprised his colleagues in biology by giving up a coveted academic position at King's College at the University of London to complete

---

9 J.S. Huxley, *Essays of a Biologist* (London, 1923).

10 J.S. Huxley, *Essays of a Humanist* (London, 1964).

11 A similar point has been made by C. Divall, 'From a Victorian to a modern: Julian Huxley and the English intellectual climate', in Waters and Van Helden (eds), *Julian Huxley* (n. 5), 31–44.

12 For a historical survey of the idea of progress and the idea of evolution, see M. Ruse, *Monad to Man* (Cambridge, MA, 1996). See also M. Nitecki (ed.), *Evolutionary Progress* (Chicago, 1988); and the sources in note 5.

what became one of his most successful and ambitious projects. He collaborated with H.G. Wells and his son 'Gip', a zoologist at University College London, on an ambitious project to lay the groundwork for *The Science of Life*, a three-volume work that appeared in 1929–30.[13] Wells and Huxley had shared a similar unified world view and a distaste for incoherence. In one well-known response to the novelist James Joyce dated 23 November 1928, Wells had expressed his frustration with the lack of coherence of Joyce's work, writing, 'Your training has been Catholic, Irish, insurrectionary, mine such as it was scientific, constructive and I suppose, English. The frame of my mind is a world wherein a big unifying and concentrating process is possible and progress not inevitable but interested and possible'. Expressing his abhorrence of Joyce's work he continued: 'Your work is an extraordinary experiment and I could go out of my way to save it from destruction or restrictive interruption. It has its believers and its following. Let them rejoice in it. To me it is a dead end'.[14] Though some correspondence exists, little of scholarly utility is known about the collaboration between Huxley and Wells on this great work outside of Julian Huxley's recollections in his *Memories*.[15] According to this source, it was a tumultuous but productive relationship that resulted in a very popular book – mostly written by Julian Huxley – that influenced biology education widely.

By the mid-1930s Huxley had also begun to connect with American colleagues pursuing similar ends, like Theodosius Dobzhansky and Ernst Mayr. In Britain he was in close contact with evolutionists like J.B.S. Haldane, C.D. Darlington along with C.H. Waddington and Joseph Needham. He was in especially close contact with R.A. Fisher and E.B. Ford. The other person who inspired Huxley was the geneticist Thomas Hunt Morgan to whom he eventually dedicated his great book on evolution entitled *Evolution: The Modern Synthesis*, which appeared in 1942.[16]

In 1936 Huxley extended his interest in evolution by supporting systematics, when he helped to organise the Association for the Study of Systematics in Relation to General Biology. This increased his interactions with other British systematists who were trying to bring consensus to the systematics community between the newer sciences of genetics and ecology and the older aims of taxonomy within a Darwinian evolutionary framework. In 1940 his efforts to launch a reform of systematics culminated with the appearance of an edited collection under the title, *The New Systematics*.[17] Suffice it to say that by the

---

13 H.G. Wells, J.S. Huxley and G.P. Wells, *The Science of Life*, 3 vols (London, 1929–30). A two-volume edition was published in New York, 1931.

14 As cited in R. Ellmann, *James Joyce: New and Revised Edition* (Oxford, 1983), 607.

15 J.S. Huxley, *Memories*, vol. 1 (London, 1970).

16 J.S. Huxley, *Evolution: The Modern Synthesis* (London, 1942).

17 J.S. Huxley (ed.), *The New Systematics* (Oxford, 1940).

mid-1930s Huxley's involvement with biologists and evolutionists along with literati and British intellectuals was extensive. If any one person in the 1930s could summarise the modern evolutionary state of the art in a palatable form for a wide readership, that person was Julian Huxley.

Evolution: the modern synthesis

The opportunity to publish a synthetic work that would draw together the burgeoning literature in evolutionary studies with his evolutionary humanism came with the request to give the presidential address to the zoology section of the British Association for the Advancement of Science in 1936.[18] In the essay written for this occasion, Huxley expressed his wish for a unified biology and observed that a move toward unification was taking place. In a now famous passage, he wrote:

> Biology at the present time is embarked upon a phase of synthesis after a period in which new disciplines were taken up in turn and worked out in comparative isolation. Nowhere is this movement towards unification more likely to be fruitful than in the many sided topic of evolution and already we are seeing its first fruits in the reanimation of Darwinism which is such a striking feature of post-war biology.[19]

This essay, entitled 'Natural Selection and Evolutionary Progress', formed the basis for Huxley's 1942 book. Entitled *Evolution: The Modern Synthesis*, it was intended to serve as a technical book that gave a status report that incorporated the new insights that had been emerging from the newer evolutionary genetics with more traditional approaches to Darwinian selection theory that accounted for the origins of biological diversity. Although it was replete with new developments from genetics, systematics and the work of mathematical population genetics – all within an ecumenical perspective on evolutionary theory – it was also a book that embodied Huxley's world view. It was in fact his most skilful weaving of science, world view and ideology, the integration of which had been long coming.

The very structure of the book revealed Huxley's intent. From chapter 1, 'The Theory of Natural Selection', which raised the status of natural selection to theory, to the final chapter, 'Evolutionary Progress', natural selection was

---

[18] Huxley first demonstrated his acumen in contemporary evolutionary theory in books 3 ('The Incontrovertible Fact of Evolution') and 4 ('The How and the Why of Development and Evolution') in Wells, Huxley and Wells, *Science of Life* (n. 13).

[19] J.S. Huxley, 'Natural Selection and Evolutionary Progress', Presidential Address at Annual Meeting, *Report of the British Association for the Advancement of Science* 106 (1936), 81–100.

used to ground a progressive vision of the world within Huxley's evolutionary humanism. The 'modern synthesis' of evolution, according to Huxley, took readers from the idea of the gene to the sociopolitical reality of the 1930s.

Beginning with a historical retelling of the period preceding the modern synthesis, Huxley described the 'eclipse of Darwin' (and coined this now famous phrase) at the turn of the twentieth century. He charted the demise of evolutionary studies, dominated by a descriptive natural history tradition, with the establishment of more experimental sciences like Mendelian genetics. The first chapter, which ostensibly was to discuss the theory of natural selection, thus became an historical account of the philosophical and methodological struggle to lend scientific legitimacy to evolution. He singled out evolution as 'the most central and most important of the problems in biology'. He also argued that selection – based on a deductive logical step for Darwin – was now a 'fact of nature capable of verification by observation and experiment'.[20] Most importantly, Huxley presented to his readers an ongoing process of integration between diverse biological disciplines leading to the unification of biology. In a revealing passage that echoed his address in 1936, Huxley argued for the unification of biology and the consequent rebirth of Darwinism:

> Biology in the last twenty years, after a period in which new disciplines were taken up in turn and worked out in comparative isolation has become a more unified science. It has embarked upon a period of synthesis, until to-day it no longer presents the spectacle of a number of semi-independent and largely contradictory sub-sciences but is coming to rival the unity of older sciences like physics, in which advance in any one branch leads almost at once to advance in all other fields, and theory and experiment march hand-in-hand. As one chief result there has been a rebirth of Darwinism.[21]

This new science of modern biology had the methodological rigour of the physical sciences, yet because it had unique processes that could not be reduced to physics and chemistry, it was also autonomous or independent of the physical sciences. Though there was plenty of room for processes like random genetic drift, Darwinian natural selection as a mechanism was responsible for evolution. The same natural selection also grounded Huxley's secular belief, his evolutionary humanism.

In choosing to ground evolutionary science and evolutionary humanism in the same process of natural selection, Huxley had to reconcile a serious contradiction. If selection were a truly mechanistic principle, then belief in evolutionary progress was exceedingly hard to sustain. If it were mechanical it was also non-teleological, and purposiveness or directionality could not be

---

20 Huxley, *Evolution: The Modern Synthesis* (n. 16), 13–28.

21 Ibid., 26.

ascribed to it. No evolutionary progress could therefore occur if there were no goal or end point for evolution, and without evolutionary progress no hope existed for human improvement. Articulating a non-teleological version of natural selection that could still somehow give direction and make possible progressive evolution, and at the same time adhering to selection as a mechanistic and therefore a legitimate scientific principle, was the challenge that he left until the very final chapter of his book.

The trick was in maintaining the balance between extremes that permitted enough room for progress. For Huxley, the handle in achieving a balance lay in the close association he made between evolutionary progress and technological progress. According to Huxley, humans had at least one ability that made them unique, and that was their ability to control their environment through technology. Through conscious, wilful use of their own evolved intelligence, humans had attained agency over their own evolution. This same intelligence would also be able to generate human values, giving rise to 'morality, pure intellect, aesthetics, and creative activity'. For Huxley, 'Man' had attained a certain status in evolutionary history by virtue of possessing this controlling intelligence. Humans were not just unlike any other animal forms, they were the most 'unique' and 'highest' of all organisms on earth.[22] For Huxley, evolution, was thus 'as much a product of blind forces as is the falling of a stone to earth or the ebb and flow of the tides', but purpose itself would only come from human will. He wrote 'if we wish to work towards a purpose for the future of man, we must formulate that purpose ourselves. Purposes in life are made, not found'.[23] Evolutionary progress in Huxley's form was as mechanistic and non-purposive as possible yet made room enough for free will.

This same evolutionary framework, which was progressive and lent human agency, also made possible the control of human heredity leading to human improvement and was translated into Huxley's ardent support of eugenics, in a 'modernised' or liberalised form that was wary of ethnic, race and class bias. Huxley's support of this 'reform' eugenics was in fact a dominant theme in his evolutionary world view.[24]

---

22 Huxley wrote an entire book on the uniqueness of humans: J.S. Huxley, *The Uniqueness of Man* (London, 1941). In the United States this book was published as *Man Stands Alone* (New York, 1941).

23 Huxley, *Evolution: The Modern Synthesis* (n. 16), 576.

24 Julian Huxley was a life-long advocate of eugenics, becoming President of the Eugenics Society in the 1960s. For an analysis of the complex twists and turns in Huxley's support of eugenics, especially his support of post-war 'reform eugenics', see G. Allen, 'Julian Huxley and the Eugenical View of Human Evolution', in Waters and Van Helden (eds), *Julian Huxley* (n. 5), 193–222; and David Hubback, 'Julian Huxley and Eugenics', in Keynes and Harrison (eds), *Evolutionary Studies* (n. 5), 194–206.

Finally, for Huxley, evolutionary progress and evolutionary humanism were to do one more thing: they would stabilise an ideology. The last two pages of *Evolution: The Modern Synthesis* reveal the complex interplay between science, world view and ideology. First echoing two of the great icons of his age, George Orwell and Eugene Ionesco, Huxley revealed his fears of the growing collective movements – of both the political right and left – of the 1930s. To Huxley, these collectives threatened to lead to 'subordination of the individual'. Upholding an overly purposive or teleological perspective was an equally distasteful prospect, since it meant that life would be fulfilled in a 'supernatural world'. The struggle to avoid both collectivism and teleology was the struggle Huxley saw facing the modern world. It was his hope that his theory of progressive evolution would provide solutions to these problems, and at the same time resolve the 'central' problem of evolution. In Huxley's theoretical formulation of progressive evolution, selection acted on the individual level so that the individual could be 'unique' at the same time that it existed in a populational or social group.[25] Neither totally mechanistic/materialistic (hence avoiding the politically extreme left wing of atheists and communists), nor too vitalistic/spiritual/mystical (hence avoiding the extreme right wing of fascists, Nazis and religious fundamentalists), Huxley's evolutionary framework balanced just enough mechanism and materialism with purpose and progress to sustain and justify a moderate liberal ideology. The threat to evolution that Huxley attributed at the end of the first chapter to the extremes of Henri Bergson, the ultravitalistic metaphysician, and William Bateson, the ultramechanistic materialist, was thus to be neutralised by a mechanistic yet purposive view of evolution. After the Second World War, Huxley and his framework would in turn sustain an increasingly moderate, popular liberal ideology with a view of an independent biological science.

Because the book had a clear political and ethical agenda and because it appeared to have an unconventional organisation (some complained it was disorganised) and perhaps because it really was not original in terms of scientific insights, it was not as well received by fellow scientists as other books that are associated closely with the evolutionary synthesis. Evolutionists never really thought the book properly synthetic in the way that Theodosius Dobzhansky's older *Genetics and the Origin of Species* had been.[26] Huxley's book did, however, become a popular account of evolution that reached far wider audiences. Huxley was a popular writer, an authority on current evolutionary science, and even more importantly, was rapidly attaining celebrity status in Britain as a result of his masterful appearances from 1939 on the BBC radio programme

25 See Huxley, *Uniqueness of Man* (n. 22).

26 T. Dobzhansky, *Genetics and the Origin of Species* (New York, 1937).

the *Brains Trust*.[27] The book appeared just before his popularity peaked as a result of this show. It was published simultaneously in Britain and the United States, was widely distributed and was cited as an authoritative source on evolution during the post-war period, especially by readers interested in general science. For these wider audiences, the book served to herald the coming of the unified biological sciences now 'modernised' and rivalling both physics and chemistry. With its emphasis on evolutionary progress, furthermore, the book offered an enquiry into an ethical system grounded in evolution and a world view that supported a liberal, secular and progressive ideology. Evolution was thus presented as a legitimate, rigorous and palatable science that was heavily materialistic and mechanistic, yet left enough room for purposiveness and free will for human beings.

The post-war period

Belief in such a scientific world view that made evolution by natural selection accountable for life became even more needed in the post-war period. As the horrors of the Holocaust became known and nuclear threat loomed with the Cold War, a framework that endorsed the fundamental adaptability of life that offered some progress, a moderate or liberal ideology, and an optimistic and coherent world view with humans as agents of their own evolution became even more urgent. For many evolutionists, only with progressive evolution would human 'improvement' be thought possible. Huxley's version of evolutionary humanism also helped account for, justify and accelerate first atomic-age and then space-age technology. By the early 1960s – the same historical moment characterised as the most prosperous and optimistic in recent American history – belief in selection culminated with panselectionism, an overarching selectionist theory supporting the view that natural selection had extreme power to shape evolutionary history: social improvement and biological improvement could both be engineered with the same evolutionary technology.

Evolutionary progress through the mechanism of evolution, though it appeared to be a contradiction, struck just the right balance between purpose, progress and mechanistic materialism – it was deterministic enough to be predictable yet not enough to remove free will – for a wide audience of evolutionists who took their views through textbooks, semi-popular works and essays addressed to the wider audience of scientists. Closely echoing Huxley were his evolutionary colleagues who upheld doctrines of evolutionary progress throughout the 1940s, 1950s and 1960s: Theodosius Dobzhansky, G. Ledyard

27 H. Thomas, *Britain's Brains Trust* (London, 1944). See also D.L. Lemahieu, 'The Ambiguity of Popularization', in Waters and Van Helden (eds), *Julian Huxley* (n. 5), 252–6.

Stebbins, Hermann J. Muller, G.G. Simpson and even Ernst Mayr.[28] All came to subscribe to versions of evolutionary progress at the same time that they made natural selection a mechanism. Each of these evolutionists in some fashion addressed the 'future of Mankind' either in concluding chapters of their books or in more popular books written at later stages of their careers. For example, G.G. Simpson strongly echoed Huxley when he wrote:

> It is another unique quality of man that he for the first time in the history of life has increasing power to choose his course and to influence his own future evolution. It would be rash indeed to attempt to predict his choice. The possibility of choice can be shown to exist. This makes rational the hope that choice may sometime lead to what is good and right for man. Responsibility for defining and for seeking that end belongs to all of us.[29]

The links between all these evolutionists were solidified, formalised and indeed *disciplined* in 1946 when they came together to found the first international Society for the Study of Evolution (SSE).[30] Huxley had been in on that from the start. In 1939, at a special 'Symposium on Speciation' organised by Dobzhansky for the American Association for the Advancement of Science (AAAS) meetings in Columbus, Ohio, Huxley met with him, Carl Epling and Ernst Mayr to suggest the formation of an official Society for the Study of Speciation. An informal society that functioned as an information service was indeed formed in the United States, but it was thwarted by the outbreak of the war. After a series of attempts to bring workers together, and after the end of war, the formal society was officially founded at the AAAS meetings in St Louis in 1946. A grant proposal submitted to the American Philosophical Society by two of the 'architects' of the synthesis, Ernst Mayr and G.G. Simpson, to help fund the new journal *Evolution*, highlighted the fact

---

28 For instance, T. Dobzhansky, *Mankind Evolving: The Evolution of the Human Species* (New Haven, 1962); idem. *The Biology of Ultimate Concern* (New York, 1967); and Dobzhansky's 'Introduction' to the 'Life Nature Library Series' volume by R.E. Moore, *Evolution* (New York, 1962). See also G.G. Simpson, *The Meaning of Evolution* (New Haven, 1949); idem, *This View of Life: The World of an Evolutionist* (New York, 1964); idem, *Biology and Man* (New York, 1969). See also an unpublished monograph in the author's possession by G. Ledyard Stebbins entitled 'Objectives and Philosophy of an Evolutionist'. And for discussion of themes of progress in Huxley's and Simpson's works, see M. Swetlitz, 'Julian Huxley, George Gaylord Simpson and the Idea of Progress in 20th Century Biology' (PhD diss., University of Chicago, 1991). For general discussion of progress in evolution, see Nitecki, *Evolutionary Progress* (n. 12), and Ruse, *Monad to Man* (n. 12).

29 Simpson, *Meaning of Evolution* (n. 28), 348.

30 V.B. Smocovitis, 'Organizing Evolution: Founding the Society for the Study of Evolution, 1939–1947', *Journal of the History of Biology* 7 (1994), 241–309.

that the journal would help in 'unifying the fields'.[31] These same individuals joined forces on the occasion of the bicentennial of Princeton University in 1947 and celebrated what they recognised as the new discipline – evolutionary biology – whose purpose would be to unify biology. In writing the summation to the edited volume of the proceedings published shortly thereafter, Hermann J. Muller wrote that a 'convergence of evolutionary disciplines' had taken place.[32] Muller argued that the process of synthesis had given rise to a new type of evolutionist, the synthetic type of evolutionist. Closing with the scientific points of agreement, Muller then moved on to explore what all this meant for the formulation of a meaningful world view. Though he recognised that a mechanistic and materialistic view of evolution did not bode well for the future of all species, he echoed Huxley in his belief that humans could rise above their own evolutionary destiny. In a soaring passage he wrote: 'If, then, we wish evolution to proceed in ways that we consider progressive, we ourselves must become the agents that make it do so. All of our studies of evolution must finally converge in that direction'.[33]

Julian Huxley had not just been a keen organiser on behalf of evolution; through his writing he had helped shape the discipline and had given it its characteristic flavour as well as its close relations to the modernised biology. He was one of the first to explicitly adopt the phrase 'evolutionary biology' in his 1942 book, giving the new science a disciplinary rubric that remained intact through much of the twentieth century. He subsequently took every opportunity to promote the new science as the central science of biology, which in turn would serve as the central science in a unified theory of scientific knowledge. In 1949, in an address entitled 'Evolution and Scientific Reality', he wrote that 'one of the outstanding events in scientific history has been the emergence, during the second quarter of the present century, of evolutionary biology as a science in its own right'. Drawing a comparison with the way that the germ theory of disease had been consumed by the new sciences of bacteriology and microbiology, Huxley then noted that 'in a rather similar way, the evolution theory has today been swallowed in that science of evolution – evolutionary biology. The difference is that, while microbiology is a departmental branch of science, involving a certain definable field, evolutionary biology is a central science, with ideas demarcating all other

---

31 As cited in J.A. Cain, 'Common Problems and Cooperative Solutions: Organizational Activities in Evolutionary Studies, 1936–1947', *Isis* 84 (1993), 1–25. Mayr and Simpson were successful in their grant application.

32 H.J. Muller, 'Redintegration of Symposium on Genetics, Paleontology and Evolution', in G. Jepsen, E. Mayr and G.G. Simpson (eds), *Genetics, Paleontology and Evolution* (Princeton, 1949; repr. New York, 1963), 421–45.

33 Ibid., 445.

branches of the life science'. Huxley continued, '[T]his you may say, is by now a commonplace. However, I do not consider that all the implications of evolutionary biology have been grasped. They have not been fully grasped at all in relation to science as a whole, from physics on the one hand to psychology and the human sciences on the other'.[34]

Nearly four years later, in another well-known book, *Evolution in Action*, Huxley extended the reach of evolutionary biology to the universe at large to make it become part of a cosmic evolutionary process:

> Evolutionary science is a discipline or subject in its own right. But it is the joint product of a number of separate branches of study and learning. Biology provides its central and largest component, but it has also received indispensable contributions from pure physics and chemistry, cosmogony and geology among the natural sciences, archaeology and prehistory, psychology and anthropology. As a result, the present is the first period in which we have been able to grasp that the universe is a process in time and to get a first glimpse of our true relation with it. We can see ourselves as history, and can see that history in its proper relation with the history of the universe as a whole.[35]

Huxley's cosmic unifying vision met with sympathetic resonance as a generation of astronomers and cosmologists began to see the universe as part of an historical process, an evolving entity that reflected progressive evolution. Cosmic, galactic, stellar, planetary, chemical, organic and cultural evolution began to emerge as a continuum in 'unified' evolutionary cosmology as astronomers, physicists, chemists and even social scientists began to look to each other, and to each other's disciplines, in their vision of a unified theory of knowledge. From leading figures like Sir James Jeans, Arthur Eddington and Fred Hoyle to Jacob Bronowski and Loren Eiseley, the belief in evolutionary progress through the mechanism of natural selection began to be extended to formulate the beginnings of a unified theory of knowledge in the post-war period. Biology occupied a central niche within this unified theory of knowledge, and immediately after the war, biology itself would be organised into one enormous 'umbrella' organisation in the United States called the American Institute of Biological Sciences (AIBS). The 'architects' of the evolutionary synthesis continued to operate within this organisation to argue that evolution unified the biological sciences. Beginning in the 1950s and continuing through the 1960s and 1970s arguments were made that

---

34 'Evolution and Scientific Reality', manuscript dated 1949, Box 67.7., Huxley Papers, Rice University, Houston, Texas.

35 J. Huxley, *Evolution in Action* (New York, 1958), 1–3.

evolution served as 'the central organising principle' of the biological sciences and that 'nothing in biology made sense except in the light of evolution'.[36]

The unity of knowledge itself – what it was, what it would entail and what it would mean for the 'future of modern man' – came to dominate discussions in wide intellectual circles in the 1950s. This was not simply a response to the philosophical Unity of Science Movement launched by the crusaders within the Vienna Circle, nor was it inspired directly by their Encyclopedia of Unified Science project; this was a new global movement that was reflected in intellectual circles worldwide as they recognised the possibility of a global community of shared values where education would play a central role. This was no more apparent than in the theme of Columbia University's bicentennial conference of 1954. The full title of the conference was 'Unity of knowledge' and the central theme was 'Man's right to knowledge and the free use thereof'. The list of contributors to and participants in this conference and its edited collection of works reflected the grand sweep of the unity of knowledge: Pierre Teilhard de Chardin, Theodosius Dobzhansky, Julian Huxley, Talcott Parsons, B.F. Skinner, Harold Urey, Niels Bohr, John Von Neumann, Willard V.O. Quine, Ernest Nagel and Philipp Frank. All of these individuals were linked, if not enveloped by the same Enlightenment ideals brought to light: unity and diversity, progress and a belief in a liberal, secular and evolutionary humanism. Huxley's own contribution to this conference once again reified and updated his grandfather's famous attempt to determine 'Man's place and role in nature'.[37]

Huxley's ideological commitments materialised most fully outside traditional academic domains, however. Not holding a traditional academic position since his resignation from King's College in 1927, Huxley put his formidable administrative talents to use in promoting science first as Secretary to the London Zoological Society between 1935 and 1942, then took the unprecedented step of becoming the first full-time secretary to United Nation's Educational and Scientific Cultural Organization (UNESCO) from 1946 to 1948. It was on his and his colleague's Joseph Needham's insistence that the 'S' for science was formally included in UNESCO. His first charge as secretary of the Preparatory Commission was to draft a charter for the new agency. It comes as no surprise that Huxley wasted no time in directing the commission; this was the perfect opportunity to apply his deepest belief in the guiding force

---

36 By 1973 this argument also warded off the rise of scientific creationist assaults in the United States aimed at doing away with the teaching of evolution in high schools. T. Dobzhansky, 'Nothing in Biology Makes Sense Except in the Light of Evolution', *American Biology Teacher* 35 (1973), 125–9.

37 See L. Leary (ed.), *The Unity of Knowledge* (New York, 1955) for a description of the conference, its aims and goals, along with a more detailed list of participants.

of evolutionary humanism. Drafting a 60-page booklet for the new commission entitled 'UNESCO, Its Purpose and Philosophy', Huxley set forth the policy that the UN could not rely on conventional religious philosophies or doctrines, nor on any academic theories. He argued forcefully for the power of 'scientific humanism', in what amounted to an evolutionary manifesto that would guide and unify humanity:

> UNESCO must work in the context of what I call *Scientific Humanism*, based on established facts of biological adaptation and advance, brought about by means of Darwinian selection, continued into the human sphere by psycho-social pressures, and leading to some kind of advance, even progress, with increased human control and conservation of the environment and natural forces.[38]

Huxley's views were met with less than enthusiasm, however. They did not sit well with other members of the commission who rejected the pamphlet, which they thought reflected a dangerously atheistic world view. A small slip was inserted into the published document stating it reflected the personal views of the author. Huxley was nonetheless given the honour the following year of being elected the first Director-General of UNESCO, though he was given the position for only two years instead of the constitutional six.[39] There is some indication that the American delegation urged the reduction of his term to the two years. Huxley served out his term and put much of his liberal philosophy to work by supporting international conservation efforts through the creation of state parks and the special preservation of natural sites. He also supported the creation of museums and institutions to safeguard cultural and historical artefacts and traditions. He was actively involved in promoting birth control, especially in developing nations. Though he had stepped down as Director-General, he continued to play an active role in drafting the first UNESCO Statement on Race, which appeared in 1950 and denied the existence of innate biological and mental differences between different human 'races'. Though some of this work has been subsequently criticised as being less than liberal, and even elitist, there is little doubt that Huxley shaped the character of the organisation during that critical early period.[40] Reflecting on Huxley's accomplishments overall, Assistant Director Richard Hoggart noted that:

---

38 J.S. Huxley, *Memories*, vol. 2 (New York, 1973), 15. See also G. Blue, 'Scientific Humanism and the Founding of UNESCO', *Comparative Criticism* 23 (2001), 173–200.

39 W.H.G. Armytage, 'The First Director-General of UNESCO', in Keynes and Harrison (eds), *Evolutionary Studies* (n. 5), 186–93. Armytage gives a fuller discussion of Huxley's election.

40 See Waters and Van Helden (eds), *Julian Huxley* (n. 5) for some of these criticisms.

> He was less perceptive in proposing that the Organization propound a world philosophy, a new humanism; that could never have been adopted. But in general he kept up the flow of good ideas through the two years of his tenure, and his speech to his last General Conference is still not only useful but exciting reading. He gave UNESCO an intellectual head of steam and a restless sense of enquiry at its start.[41]

Walter Laves and Charles Thomson, respectively the assistant director of UNESCO and the director of the US National Commission of UNESCO under Huxley, echoed some of Hoggart's reflections when they wrote, 'Probably no one person more directly influenced the content and direction of UNESCO. Indeed, he was largely responsible for charting the broad course to which the organisation became committed during the early years'.[42] Whatever the influence of Huxley's two years in UNESCO, the 'S' survives and serves as a permanent reminder of Huxley's guiding mission.

Huxley's unifying vision extended

After leaving his position at UNESCO, Huxley continued his work on related commissions promoting his liberal causes and continued to capitalise on his fame and celebrity status by writing for mostly popular or semi-popular audiences. His 1942 book continued to be read and cited widely. He travelled extensively, continued to make public appearances and, inevitably, found newer and more popular audiences that were opened up and created by the growing technology of telecommunications. Capitalising on the celebrity he had acquired in the 1940s through the *Brains Trust*, one of the most popular radio programmes of the BBC, Huxley continued to charm audiences throughout the 1950s and 1960s as he was frequently heard as spokesperson of science, first on radio and then on television programmes. He was an engaging speaker and a charismatic showman.

This native talent and his world view were put to use at what we can view as the apogee of the evolutionary synthesis when Huxley was invited to give a special convocation address in 1959 on the occasion of the 100th anniversary of the publication of Darwin's *On the Origin of Species*.[43] A grand celebration organised by anthropologist Sol Tax and a special committee at the University

---

41 Richard Hoggart, *An Idea and its Servants: UNESCO from Within* (Oxford, 1978).

42 W.H.C. Laves and C.A. Thomson, *UNESCO: Purpose, Progress, Prospects* (Bloomington, 1957), 295.

43 V.B. Smocovitis, 'The 1959 Darwin Centennial Celebration in America', in Clark Elliott and Pnina Abir-Am (eds), *Commemorations of Scientific Grandeur*, *Osiris* 14 (1999), 274–323.

of Chicago was staged that included an organised media blitz rivalling that of the coronation of Queen Elizabeth II. As grandson of Thomas Henry, and as honorary visiting professor at the University of Chicago for the year, Huxley was one of the highlights of the five-day conference that culminated in a special secular service on Thanksgiving Day in the Rockefeller Chapel. Given the pulpit for the convocation address, Huxley delivered what is probably his most famous – indeed notorious – lecture of all, his 'secular sermon' entitled 'The Evolutionary Vision' (Figure 3.1). Huxley's lecture once again embodied his unifying vision made possible by evolutionary humanism. Religion, according to Huxley, was but an 'organ of evolving man', and the time had come for a 'new organization of thought' based on the new evolutionary vision. He continued:

> Evolutionary man can no longer take refuge from his loneliness in the arms of a divinized father-figure whom he has himself created, nor escape from the responsibility of making decisions by sheltering under the umbrella of Divine Authority, nor absolve himself from the hard task of meeting his present problems and planning his future by relying on the will of an omniscient, but unfortunately inscrutable, Providence.

Furthermore, the 'evolutionary vision,' as 'opened up to us' by Charles Darwin, 'exemplifies the truth that truth is great and will prevail, and the greater truth that truth will set us free'. 'Evolutionary truth' could, he added, free humans from 'subservient' fears, and showed the way to 'our destiny and our duty'.[44]

On that occasion, at least, Huxley's 'secular sermon' was not well received: many of the audience members, including numerous scientists, later reported that they were stunned by the lack of discretion that Huxley had displayed, though none so publicly or vociferously as the popular press which feasted on the newsworthiness of Huxley's fire and brimstone secular sermon. Clearly not all was acceptable in Huxley's ideology, especially when stated this baldly to conservative American audiences. Yet Huxley's evolutionary humanism, in various guises, continued to have broad enough and, I would argue, enduring appeal throughout the 1950s and well into the closing decades of the twentieth century. Versions of it were picked up and extended, if not echoed, by other popularisers at the same time, such as Loren Eiseley. His popular book *Immense Journey* picked up much the same cosmic striving for a unified whole based on evolution and influenced a generation of younger scientists.[45] Evolutionary humanism was perhaps best seen in the world

---

[44] J. Huxley, 'The Evolutionary Vision: The Convocation Address', in Sol Tax and Charles Callender (eds), *Evolution after Darwin*, vol. 3 (Chicago, 1960), 249–61.

[45] L. Eiseley, *The Immense Journey* (New York, 1959).

Figure 3.1 Sir Julian Huxley at the pulpit of the Rockefeller Chapel, Chicago, 1959, delivering his address 'The Evolutionary Vision'.

Source: Darwin Centennial Papers, Department of Special Collections, Joseph Regenstein Library, University of Chicago. Reproduced with permission.

view transmitted by Jacob Bronowski in his popular television show *The Ascent of Man*, in the early 1970s.[46] And, Huxley's evolutionary humanism reached vast international television audiences in the 1980s and well into

46 J. Bronowski, *The Ascent of Man* (London, 1973).

the 1990s through the American Public Broadcasting Service show *Cosmos*.[47] Beginning with episode 2, entitled 'One Voice in the Cosmic Fugue', Carl Sagan grounded his unifying vision of the cosmos in evolutionary theory as it had emerged from the work of Julian Huxley and the other architects of the evolutionary synthesis.[48]

Carrying the torch of unity and synthesis into the twenty-first century was the unchallenged heir to the synthesis, E.O. Wilson. His unifying project extended the synthetic theory to realms of human social behaviour, beginning with his controversial 1975 book *Sociobiology: The New Synthesis*.[49] A compact popular account, *On Human Nature*,[50] extended his evolutionary vision to include humanistic concerns like agency, free will and religious belief. Wilson's book *Consilience: The Unity of Knowledge*, the most ambitious of all, argued for the extension of the synthetic project to the whole of knowledge, bringing the humanities into the realm of the sciences.[51] Through what he terms 'unification metaphysics', Wilson claims release from 'the confinement of fundamentalist religion'. The concluding chapter of this book echoes beliefs that all these books have in common and that originated with Huxley: although humans are animals subject to mechanistic forces like natural selection, humans are unique in their ability to know and to override evolutionary determinism. According to Wilson, 'thanks to advances in genetics and molecular biology, hereditary change will soon depend less on natural selection than on social science'. The prospect of what he terms 'volitional evolution', which Wilson takes to mean 'a species deciding what to do about its own heredity', will present 'the most profound intellectual and ethical choices that humanity has ever faced'.[52] Evocative of Huxley he concludes that *Homo sapiens* is the first 'truly free species', because it is the species that 'decommissions' natural selection as the 'force that made us'.[53]

---

47 C. Sagan, *Cosmos* (New York, 1980).

48 Sagan had once been a student of H.J. Muller, who in turn had worked closely with Julian Huxley.

49 E.O. Wilson, *Sociobiology: The New Synthesis* (Cambridge, 1975).

50 E.O. Wilson, *On Human Nature* (Cambridge, 1978).

51 E.O. Wilson, *Consilience: The Unity of Knowledge* (New York, 1998).

52 Ibid., 273.

53 Ibid, 276. For more discussion on evolutionary humanism, the evolutionary synthesis and its influence in biological anthropology see V. B. Smocovitis, 'Humanizing Evolution: Anthropology, the Evolutionary Synthesis, and the Prehistory of Biological Anthropology, 1927–1962', *Current Anthropology* 53 (2012), 108–25.

Closing thoughts

Evolutionary humanism, thanks first to the special efforts of biologists like Huxley, other architects of the evolutionary synthesis and their numerous disciples and heirs like Carl Sagan and E.O. Wilson, reached vast popular audiences in the latter half of the twentieth century. It was also simultaneously disciplined through the efforts of a narrower community of evolutionists through the science of evolutionary biology and its organisations like the Society for the Study of Evolution. These beliefs were extended through textbooks, semi-popular books, radio and television programmes to vast audiences. How has the drive for unity affected the efforts of such scientists? In the case of evolution I would argue that it has been the central problem in the field; the 'unification metaphysics' described by Wilson and so perfectly demonstrated by his attempts to unify all knowledge in *Consilience* is probably held up by a majority of practising evolutionary biologists. As the case of Wilson also demonstrates, some version of Huxley's evolutionary humanism remains alive and well within a liberal, secular framework, even though evolutionary progress is no longer a credible belief in evolutionary science.

What general conclusions can we then draw about the attempts to unify biology in the twentieth century? For one thing, we see how evolution served as the vehicle for the unification of the biological sciences following the evolutionary synthesis. Arguments like 'evolution is the central organising principle' and 'nothing in biology makes sense except in the light of evolution' abound in the period following the evolutionary synthesis. Within what had previously been perceived as the fragmented biological sciences, evolution was to serve as the unifying element of a unified modern biology. It simultaneously performed a range of functions that led to arguments for not just the unity of biology, but also for the unity of the sciences, the unity of knowledge and all ultimately offering a coherent world view. This world view offered an ethical or moral system, a stable political ideology – liberal, progressive and secular – at the same time that it supported an autonomous science of biology. The appeal of this world view is perhaps due to the fact that, being grounded in evolutionary science, it appeared benignly political or even apolitical. For Huxley himself it was a way to truth, which would prevail in the service of universal 'mankind'. Such is the power of evolutionary science, which could appear, and I would argue continues to appear, as though it were a value-neutral science capable of supporting a world view that is taken to incorporate universal human values and to be ideologically neutral.

Acknowledgements

I wish to thank Harmke Kamminga and Geert Somsen for their helpful assistance with this manuscript. Thanks to Thomas Söderqvist for critical commentary.

Chapter 4

Unity through experiment?

Reductionism, rhetoric and the politics of nuclear science, 1918–40

Jeff Hughes

'Science', Ernest Rutherford famously proclaimed, 'is either physics or stamp-collecting'.[1] For Rutherford, the leading light both of interwar nuclear science and of British and imperial science more generally, 'physics' was the study of atoms, nuclei and the even more basic constituents of matter. Up against one of Rutherford's own 'bright young men' from the Cavendish Laboratory at an interview for a Cambridge research studentship in 1930, the young C.P. Snow, then embarking on a career in spectroscopic research, found that one of his interviewers was Rutherford himself. Snow recalled that Rutherford 'took out his pipe and turned on to me an eye which was blue, cold and bored. . . . Wasn't his man, and a very good man, in for this job? What was this other fellow doing here? Why were we all wasting our time?' Rutherford, Snow went on, 'asked me one or two questions in an irritated impatient voice. What was my present work? What could spectroscopy tell us anyway? Wasn't it just "putting things into boxes"?' As 'cheerfully as [he] could manage', Snow 'asked if he couldn't put up with a few of us not doing nuclear physics' and put his case for spectroscopy as best he could.[2] Snow did not win the studentship, but clearly long remembered this episode.[3]

Rutherford's insistence on the centrality of his own reductionist form of experimental physics was more than hubris. Against the mathematical and metaphysical speculations of theoreticians and philosophers in the new era of relativity in the early 1920s, the experimental elaboration of the basic building blocks of matter and their interrelations would provide the underpinnings of a fundamental and irreducibly material understanding of the properties and behaviour of matter at micro and macro levels. Where the theorists disputed time, space and ether, airing troubling disputes about the philosophical

---

1 Quoted in J.D. Bernal, *The Social Function of Science* (London, 1939), 9.

2 C.P. Snow, *Variety of Men* (London, 1967), 3.

3 On Snow's abortive scientific career, see J. Halperin, *C.P. Snow: An Oral Biography* (London, 1983), 93–4; J.C. Brand, 'The Scientific Papers of C.P. Snow', *History of Science* 26 (1988), 111–27; N. Tredell, *C.P. Snow: The Dynamics of Hope* (Basingstoke, 2012).

foundations of science and employing the language of 'revolution' with the effect – intentional or otherwise – of undermining its historical stability, this experimentally grounded reductionist regime of research would provide the grounds for unifying agreement on the bedrock facts of matter. Rutherford's programme at the Cavendish Laboratory was not uncontested: challenges came both from other laboratories working in the fields of radioactivity and nuclear physics, and from physics laboratories specialising in other fields which represented different forms of practice, epistemology and professional values. Yet the appeal of reductionism commanded scientific and popular attention in the interwar years. Not only scientific texts but scores of more accessible books, newspaper and magazine articles, radio programmes, public displays and museum exhibitions disseminated the view that the physics of atoms and nuclei offered a new and foundational understanding of matter.[4] Beginning with tabletop nuclear disintegrations and ending with large atom-smashing machines that were laboratories in their own right, reductionist physics – a physics built on the ideology that the behaviour of matter could ultimately be explained by the properties of its 'fundamental' constituents – became established institutionally and intellectually in the interwar years.[5] It shaped both the material development of physics and intellectual expectations of science more generally for the rest of the twentieth century, and after.[6]

The nature of this foundational period in the modern regime of microphysics has, however, often been misrepresented and misunderstood. Often seen as a 'golden age' of small-scale physics when sealing wax and string reigned supreme and when internationalist values and the joy of abstract truth seeking in reductionist forms of practice were self-evident, the canonical history of interwar nuclear physics has been the victim of a particularly acute form of retrospection which draws heavily on the much later reminiscences of practitioners, who created an idealised image of physics before Big Science

---

[4] P. Bowler, *Science for All: The Popularization of Science in Early Twentieth-Century Britain* (Chicago, 2009).

[5] P. Galison, *Image and Logic: A Material Culture of Microphysics* (Chicago, 1997); J. Hughes, *The Manhattan Project: Big Science and the Atom Bomb* (Cambridge, 2003).

[6] On wider disciplinary impacts, see, for example, K. Gavroglu and A. Simoes, *Neither Physics nor Chemistry: A History of Quantum Chemistry* (Cambridge, MA, 2012); E.F. Keller, 'Physics and the Emergence of Molecular Biology: A History of Cognitive and Political Synergy', *Journal of the History of Biology* 23 (1990), 389–409; L.E. Kay, 'Quanta of Life: Atomic Physics and the Reincarnation of Phage', *History and Philosophy of the Life Sciences* 14 (1992), 3–21; P. Abir-Am, 'The Molecular Transformation of Twentieth-Century Biology', in J. Krige and D. Pestre (eds), *Science in the Twentieth Century* (Amsterdam, 1997), 495–524; S. de Chadarevian, *Designs for Life: Molecular Biology after World War II* (Cambridge, 2002).

and the Bomb.[7] Selectively emphasising the development of theory, this account has made reductionist physics benignly self-evident, linear and teleological, if paradoxically taking as its key historical moments those events that later played an important part in the development of the atomic bomb. This classically internalist 'bomb historiography' of interwar nuclear physics has been complemented by an 'external' history which also drew heavily on actors' own accounts and value judgements, and which projected backwards into the interwar period a set of values which were as often imagined as real.[8] Canonical sites (such as Bohr's Institute for Theoretical Physics in Copenhagen, the Cavendish Laboratory in Cambridge and Max Born's Physical Institute in Göttingen) and institutions (such as the Solvay Congresses, which acted as the institutional midwives to the birth of quantum mechanics, and the Nobel Prizes, with their apparently internationalist and objective character), have come to be seen through this 'external' history as exemplary indicators of the self-evident value of interwar nuclear physics.[9]

Clearly, the significance of the reductionist regime has been too easily assumed and too rarely questioned, and its *contemporaneous* scientific and cultural significance have been largely overlooked. After the Great War, when science in Britain increasingly came to be equated with industrial and military developments – exemplified not least by media hype about 'death rays' in 1924[10] – Rutherford and his co-workers had to work hard to promote atomic and nuclear science as the most socially disengaged of 'pure' sciences, as epistemologically 'fundamental' and as capable of ultimately unifying the sciences both ideologically, through its implicit transnationalist character, and

---

7 This critique of the historiography of nuclear physics is developed further in J. Hughes, 'Radioactivity and Nuclear Physics', in M.J. Nye (ed.), *The Cambridge History of Science*, Vol. 5: *The Modern Physical and Mathematical Sciences* (Cambridge, 2002), 350–74. Also see R. Staley, 'Trajectories in the History and Historiography of Physics in the Twentieth Century', *History of Science* 51 (2013), 151–77.

8 C. Weiner, 'Institutional Settings for Scientific Change: Episodes from the History of Nuclear Physics', in A. Thackray and E. Mendelsohn (eds), *Science and Values: Patterns of Tradition and Change* (New York, 1974), 187–212; C. Weiner (ed), *Exploring the History of Nuclear Physics* (New York, 1972); R.H. Stuewer (ed.), *Nuclear Physics in Retrospect: Proceedings of a Symposium on the 1930s* (Minneapolis, 1979).

9 J. Mehra, *The Solvay Conferences on Physics: Aspects of the Development of Physics since 1911* (Dordrecht, 1975); E. Crawford, *Nationalism and Internationalism in Science, 1880–1939: Four Studies of the Nobel Population* (Cambridge, 1992). Recent, more contextual and critical perspectives can be found in P. Marage and G. Wallenborn (eds), *The Solvay Councils and the Birth of Modern Physics* (Basel, 1999); R.M. Friedman, *The Politics of Excellence: Behind the Nobel Prizes* (New York, 2001).

10 J. Foster, *The Death Ray: The Secret Life of Harry Grindell Matthews* (n.p., 2009).

cognitively, through its revelation of the underlying and putatively universal constituents of matter. Though rarely stated explicitly, still less programmatically, this reductionist dogma presupposed a specific methodology. Rejecting the 'speculation' and metaphysical idealism associated with public discussions of relativity on the one hand and later the scientific utilitarianism associated with interwar Marxism and militarism on the other, Rutherford instead insisted on careful, dispassionate and disciplined experiment as the ground for proper scientific practice and warranted scientific knowledge. From the secure knowledge so produced, the essential 'building blocks of matter' would be uncovered and the foundations laid for an understanding of the construction of the house of nature.

Rutherford and like-minded reductionists promoted their ideas and the experimental ideology of their programme tirelessly not only within scientific circles, but also in the public sphere, where they portrayed 'modern' physics – the physics of atoms, nuclei and (sometimes) quanta – as the key science underpinning all other branches of scientific enquiry. Both professional and popular audiences responded positively to the message, embracing reductionism and the unification it ultimately promised, almost to the point that it became a self-evident article of faith in public intellectual culture. Yet behind the public face of optimism and projected unification, all was not well with the reductionists' experimental programme. Originally confined to one British laboratory in the early 1920s, the programme seemed to be achieving its aims through controlled, disciplined laboratory work – and careful public presentation of that work. With the entry of other laboratories into the field from the mid-1920s, however, controversies arose which threatened publicly to undermine the unifying epistemological and ideological claims of the programme. The building blocks of matter seemed to fit together in different ways in different laboratories. Against a public rhetoric of benign scientific internationalism, fierce private competition and controversy between laboratories became a powerful driving force in the development of atomic and nuclear science.

For the reductionists, controversy threatened not only the epistemological promise of their project but its credibility as a force for unification and as a model for scientific enquiry more generally. In this context, scientific unity and the public authority of the reductionist programme had to be fought for by the creation of consensus through private appeals to *social* and *professional* unity. In this chapter, I explore some of these disputes and the outcomes of this clash between ideologies, rhetoric and practices. Focussing mainly on British nuclear physicists, and particularly on Rutherford, their leader and chief public representative, I try to show how such 'unity' as there was in practice was guaranteed not by nature, method or even language, but by disciplinary politics and by the increasing availability of standardised commercial instrumentation from large electrical companies. In the end, I suggest, the experimentalists most closely engaged with the quest for 'unity' retreated from any strong claims about the reductive and unifying force of their work, and it was others *outside* that immediate circle who most obstinately clung to the idea of reductionism as a unifying force – with lasting implications for its historiography.

Radioactivity, the 'radioactivists' and the 'Heroic Age' of nuclear science

Just before Ernest Rutherford was appointed Director of Cambridge's Cavendish Laboratory in April 1919, Joseph Larmor, Lucasian Professor of Mathematics, one of the electors to the Cavendish chair and a long-standing admirer of Rutherford's radioactivity research at Manchester University, had told him that his task would be to help make Cambridge 'the Imperial University that it is expected to be in the new scheme of things'.[11] In his first post as Professor of Physics at McGill University, Montreal, Rutherford had made significant contributions to the analysis of the radiations from radioactive substances, and, with the English chemist Frederick Soddy, developed the disintegration theory of radioactivity. For this work, Rutherford won the 1908 Nobel Chemistry Prize, establishing his international reputation.[12] During his 12 years at Manchester, Rutherford had made his laboratory the centre of a small but vibrant international network of institutions in Paris, Berlin, Vienna and elsewhere, defined by the constitutive presence of rare and expensive radioactive materials as research objects.[13] Within this network, the key researchers knew each other personally in a relatively tightly knit community that one of them (James Chadwick) would later label the 'radioactivists'.[14] Mediated by judgements of scientific expertise

---

11 J. Larmor to Rutherford, 4 March 1919, quoted in D. Wilson, *Rutherford: Simple Genius* (London, 1983), 408.

12 T.J. Trenn, *The Self-Splitting Atom: A History of the Rutherford-Soddy Collaboration* (London, 1977). On Rutherford's early career, see Wilson, *Rutherford* (n. 11); J. Campbell, *Rutherford: Scientist Supreme* (Christchurch, New Zealand, 1999); M. Bunge and W.R. Shea (eds), *Rutherford and Physics at the Turn of the Century* (New York, 1979); K. Dean, 'Inscribing Settler Science: Ernest Rutherford, Thomas Laby and the Making of Careers in Physics', *History of Science* 41 (2003), 217–40.

13 M. Rentetzi, *Trafficking Materials and Gendered Experimental Practices: Radium Research in Early 20th Century Vienna* (New York, 2008); M.C. Malley, *Radioactivity: The History of a Mysterious Science* (Oxford, 2011).

14 See J. Hughes, 'The Radioactivists: Community, Controversy and the Rise of Nuclear Physics' (PhD diss., University of Cambridge, 1993); S. Boudia and X. Roqué (eds), 'Science, Medicine and Industry: The Curie and Joliot-Curie Laboratories', special issue of *History and Technology* 13 (1997).

and the material culture of radioactive substances, this network sustained an epistemic community embracing researchers affiliated in more conventional disciplinary and institutional terms to a variety of sciences, including chemistry, physics, geology and medicine.[15]

While some institutions (such as Marie Curie's Paris laboratory) had devoted their attention to the chemical aspects of radioactivity, Rutherford had developed the experimental microphysics of matter imbibed during his apprenticeship under J.J. Thomson at the Cavendish Laboratory in the late 1890s.[16] At Manchester from 1907, Rutherford turned his attention increasingly to questions of atomic structure and the physical interpretation of the increasingly varied and bewildering phenomena of radioactivity.[17] It was at Manchester that he and his growing research group pioneered the technique of using the energetic particles emitted in radioactive decay as projectiles. These projectiles were used to 'bombard' various substances, the results of the interaction being detected by various means, and the results deployed to make inferences about the composition – soon increasingly understood as the *structure* – of the bombarded material. It was out of such experiments that Rutherford had developed the nuclear theory of the atom in 1911.[18] Pitched against other atomic theories such as those of Thomson or the London mathematical physicist John William Nicholson, Rutherford's theory at first had little conspicuous success.[19]

---

[15] L. Badash, *Radioactivity in America: Growth and Decay of a Science* (Baltimore, 1979); Rentetzi, *Trafficking Materials* (n. 13); M. Lavine, *The First Atomic Age: Scientists, Radiations, and the American Public, 1895–1945* (New York, 2013).

[16] On Thomson and the training of researchers at the Cavendish Laboratory, see I. Falconer, 'J.J. Thomson and "Cavendish physics"', in F.A.J.L. James (ed.), *The Development of the Laboratory: Essays on the Place of Experiment in Industrial Civilisation* (London, 1989), 104–17; D.-W. Kim, *Leadership and Creativity: A History of the Cavendish Laboratory, 1871–1919* (Dordrecht, 2002). For a range of historical approaches to Thomson's work on the 'corpuscle' of negative electricity subsequently known as the 'electron', see J.Z. Buchwald and A. Warwick (eds), *Histories of the Electron: The Birth of Microphysics* (Cambridge, MA, 2001); T. Arabatzis, *Representing Electrons: A Biographical Approach to Theoretical Entities* (Chicago, 2006); J. Navarro, *A History of the Electron: J.J. and G.P. Thomson* (Cambridge, 2012).

[17] For Rutherford at Manchester, see Wilson, *Rutherford* (n. 11), 216–405; J.B. Birks (ed.), *Rutherford at Manchester* (London, 1962); J. Hughes, 'Report: William Kay, Samuel Devons and Memories of Practice in Rutherford's Manchester Laboratory', *Notes and Records of the Royal Society* 62 (2008), 97–121.

[18] J.L. Heilbron, 'The scattering of α and β particles and Rutherford's atom', *Archive for History of Exact Sciences* 4 (1967–68), 247–307.

[19] I. Falconer, 'J.J. Thomson's Work on Positive Rays, 1906–1914', *Historical Studies in the Physical Sciences* 18 (1988), 265–310; R. McCormmach, 'The Atomic Theory of John William Nicholson', *Archive for History of Exact Sciences* 3 (1967–68), 160–84.

It was mainly through its modification by the theoretical physicists Niels Bohr and Arnold Sommerfeld, who linked it with the quantum theory and used it to offer explanations of a number of physical and chemical phenomena, such as spectroscopic data, that Rutherford's atom came to prominence as a plausible – if still contested – account of atomic constitution.[20]

In a series of experiments carried out in 1917, between stints of war work for the British Admiralty, Rutherford had seemingly found that hydrogen nuclei – particles he would later christen 'protons' – could apparently be ejected from nitrogen under bombardment by energetic alpha particles. This finding that the nucleus could be 'disintegrated' not only supported the nuclear hypothesis, but also opened up the possibility that the nuclei of the different elements might have internal structure susceptible to experimental investigation. The first results of this work appeared in print in 1919, shortly before Rutherford arrived at Cambridge.[21] By this time, however, the post-war political and public intellectual context was creating new problems of legitimation for the emerging science of the nucleus. In the troubled politics of knowledge after the war, disillusionment with science, technology, mechanisation and rationalisation – and sometimes its flipside, a rejuvenated, almost utopian faith in the same forces – figured centrally in debates about social and moral values and national regeneration.[22] Unlike his erstwhile colleague Soddy, whose popular 1909 book *The Interpretation of Radium* (published in a new edition in 1920) had played up the potential of radioactivity as a source of energy and of a utopian future, Rutherford generally took a conservative view of the transformative potential of his subject.[23] On the one hand, he now stressed even more than before the

---

[20] J.L. Heilbron and T.S. Kuhn, 'The Genesis of the Bohr Atom', *Historical Studies in the Physical Sciences* 1 (1969), 211–90; S. Seth, *Crafting the Quantum: Arnold Sommerfeld and the Practice of Theory, 1890–1926* (Cambridge, MA, 2010); H. Kragh, *Niels Bohr and the Quantum Atom: The Bohr Model of Atomic Structure, 1913–1925* (Oxford, 2012).

[21] Wilson, *Rutherford* (n. 11), 386–405.

[22] C. Lawrence and A.-K. Mayer, 'Regenerating England: An Introduction', in C. Lawrence and A.-K. Mayer (eds), *Regenerating England: Science, Medicine and Culture in Interwar Britain* (Amsterdam, 2000), 1–23; C. Carson, A. Kojevnokov and H. Trischler (eds), *Weimar Culture and Quantum Mechanics: Selected Papers by Paul Forman and Contemporary Perspectives on the Forman Thesis* (London, 2011); R. Lettevall, G. Somsen and S. Widmalm (eds), *Neutrality in Twentieth-Century Europe: Intersections of Science, Culture and Politics after the First World War* (New York, 2012).

[23] F. Soddy, *The Interpretation of Radium: Being the Substance of Six Free Popular Lectures Delivered at the University of Glasgow, 1908* (London, 1909), revised as *The Interpretation of Radium and the Structure of the Atom* (London, 1920). The book provided fuel for H.G. Wells's futurist imaginings, resulting in the appearance of his 1914 novel *The War in the Air*, which was dedicated to Soddy and which has been widely (and anachronistically) read as a prophecy of nuclear weapons. See L. Merricks, *The World Made New: Frederick*

'purity' of his work and its 'fundamental' nature, thus emphasising its distance from the military science on which he and many others had worked during the war. At the same time, however, Rutherford sought to distance radioactivity and nuclear science from the disturbing scientific 'revolutions' associated with the much-publicised 'proof' of the theory of relativity in 1919 – establishing a scepticism towards abstract mathematical theory that would last until the end of the decade.[24] In attempting to position radioactivity in the modernism/anti-modernism debate in such a way as to neutralise its connection both with the overtly utilitarian applications of science and with the pointedly abstract nature of high (and potentially revolutionary) theory, Rutherford now sought to create a new intellectual space for radioactivity and nuclear science.[25] In all his public pronouncements aimed at managing public discourse about radioactivity in the 1920s, he grounded new developments firmly and explicitly in *experiment.*

Rutherford was responding to significant changes in the post-war disciplinary landscape, too. Though he re-energised his correspondence with the German and Austrian radioactivists in 1919, the economic difficulties in which they found themselves effectively debarred them from active research. This situation allowed Rutherford to acquire on a permanent basis the radium formerly loaned to him by the Austrian Academy of Sciences, a substantial asset to his laboratory's experimental programme.[26] The retirement in 1912 (and death in 1916) of the London chemist William Ramsay had removed one of Rutherford's principal English adversaries, while Soddy, appointed Professor of Inorganic Chemistry at

---

Soddy, Science, Politics, and Environment (Oxford, 1996), 67–9; D.C. Smith, *H.G. Wells: Desperately Mortal* (New Haven, 1986), 83–4. More generally on Soddy, see the useful essays in G.B. Kauffman (ed.), *Frederick Soddy (1877–1956): Early Pioneer in Radiochemistry* (Dordrecht, 1986).

24 L. Badash, 'Ernest Rutherford and Theoretical Physics', in R. Kargon and P. Achinstein (eds), *Kelvin's Baltimore Lectures and Modern Theoretical Physics* (Cambridge, MA, 1987), 349–73. For the 1919 eclipse expeditions and the 'proof' of relativity, see J. Earman and C. Glymour, 'Relativity and Eclipses: The British Eclipse Expeditions of 1919 and their Predecessors', *Historical Studies in the Physical Sciences* 11 (1980), 49–85; A. Sponsel, 'Constructing a "Revolution in Science": The Campaign to Promote a Favourable Reception for the 1919 Solar Eclipse Experiments', *British Journal for the History of Science* 35 (2002), 439–67; M. Stanley, *Practical Mystic: Religion, Science, and A.S. Eddington* (Chicago, 2007).

25 On post-war discussions of science (especially relativity) and modernity, see D. Bradshaw, 'The Best of Companions: J.W.N. Sullivan, Aldous Huxley and the New Physics', *Review of English Studies* 47 (1996), 188–206, 352–62; M. Whitworth, *Einstein's Wake: Relativity, Metaphor, and Modernist Literature* (Oxford, 2001); K. Price, *Loving Faster than Light: Romance and Readers in Einstein's Universe* (Chicago, 2012).

26 Wilson, Rutherford (n. 11), 436.

Oxford in 1919, became involved in institutional politics and issues of social and financial reform, and did not contribute significantly to research in radioactivity thereafter.[27] Rutherford and his Cambridge students and colleagues therefore now had the subject of radioactivity, and particularly the new question of nuclear disintegration, largely to themselves.

Against this background, the Cavendish researchers set to work. Building on the results from experiments using several different kinds of apparatus – cloud chambers, mass spectrographs, spectroscopes, ionisation chambers and, above all, scintillation counters – Rutherford and his co-workers gradually began to map the characteristics of the nuclear atom and the 'architecture of matter'.[28] Helped by the fact that they now worked in a largely uncontested disciplinary space, by 1922 they had, it seemed, successfully charted various elements of nuclear structure and were attracting increasing attention from the wider scientific community.[29] Rutherford and his allies promoted the emerging ideas of nuclear structure assiduously in the early 1920s in presentations, books and popular articles. It is often forgotten, for example, that Rutherford succeeded Thomson not only as Director of the Cavendish Laboratory but also, in 1922, as Professor of Natural Philosophy at the Royal Institution (RI) in London. Here he devoted his annual series of lectures to his laboratory's latest research on the nucleus, presenting it as a self-evidently significant and successful

---

27 Ramsay, Professor of Chemistry at University College London from 1887 to 1912, had entered radioactivity research in the early 1900s and produced work that was not regarded as credible by the community of radioactivists, and which Rutherford (and others) largely dismissed as scientific grandstanding. See T.J. Trenn, 'The Justification of Transmutation: Speculations of Ramsay and Experiments of Rutherford', *Ambix* 21 (1974), 53–77; M. Morrison, *Modern Alchemy: Occultism and the Emergence of Atomic Theory* (Oxford, 2007), 115–34. On Soddy, see J.B. Morrell, *Science at Oxford 1914–1939: Transforming an Arts University* (Oxford, 1997), 320–26; Merricks, *The World Made New* (n. 23), 84–132.

28 A.J. Lotka, 'The Architecture of Matter', *Harper's Monthly Magazine* 140 (1920), 679–87.

29 On the work and instrumentarium of the Cavendish in the early 1920s, see J. Hughes, 'Making Isotopes Matter: Francis Aston and the Mass-Spectrograph', *Dynamis* 29 (2009), 131–65; idem, 'Plasticine and Valves: Industry, Instrumentation and the Emergence of Nuclear Physics', in J.P. Gaudillière and I. Löwy (eds), *The Invisible Industrialist: Manufactures and the Construction of Scientific Knowledge* (London, 1998), 58–101. On the parallel development of nuclear models by Rutherford and others, see the key studies by R.H. Stuewer: 'The Nuclear Electron Hypothesis', in W.R. Shea (ed.), *Otto Hahn and the Rise of Nuclear Physics* (Dordrecht, 1983), 19–67; 'Rutherford's Satellite Model of the Nucleus', *Historical Studies in the Physical Sciences* 16 (1986), 321–52; 'Gamow's Theory of Alpha Decay', in E. Ullmann-Margalit (ed.), *The Kaleidoscope of Science* (Dordrecht, 1986), 147–86.

programme. These lectures were widely reported in both general scientific and engineering journals and in the press, and were therefore an excellent public venue to promote the work of the Cavendish. After one of Rutherford's series of lectures at the RI in 1922, for example, *Nature* editorialised in glowing terms:

> We think it no exaggeration to say that these experiments are some of the most fundamental which have ever been made. It is not often that a scientific discovery excites interest outside the narrow circle of the laboratory or the scientific lecture-room. . . . So fundamental are the consequences of this new discovery that the intellectual world at large must follow with keenest interest the progress of the experiments associated with the name of Rutherford.[30]

If *Nature* represented a general scientific constituency and could be expected to bang the establishment drum for Rutherford and his programme, there was wider public coverage and interest, too. In the press, the atom could be as mysterious as James Jeans's universe, and the 'private life of the atom' could be laid bare by the 'probing eyes of science' in 'Sir Ernest Rutherford's new billiards game'.[31] While the disintegration of atoms still excited hopes of a cheap new source of energy – a view consistently played down by Rutherford himself – the press took on board and pressed home the message that 'electrical atoms' were the 'one fundamental thing' in 'nature's workshop' – the 'bricks' out of which 'Nature fashions everything that exists in the universe.'[32]

This public articulation of Cantabrigian experimental reductionism reached its apotheosis when Rutherford took the Presidential Chair of the British Association for the Advancement of Science (BAAS) at Liverpool in September 1923. In a Presidential Address widely reported in the press (including a full-page spread in the *Illustrated London News*) and broadcast nationally to hundreds of thousands of listeners by the fledgling British Broadcasting Company, Rutherford triumphantly articulated the progress of his reductionist regime:

---

30 'Contemporary Alchemy', *Nature* 109 (1922), 601–2, on 602. On the RI in this period, see J. Hughes, 'Craftsmanship and Social Service: W.H. Bragg and the Modern Royal Institution', in F.A.J.L. James (ed.), *'The Common Purposes of Life': Science and Society at the Royal Institution of Great Britain* (London, 2002), 225–47.

31 'Mysteries of the Atom. The Disintegration of an Element', *The Times*, 5 April 1924, 8; 'The Private Life of the Atom. Probing Eyes of Science. Sir Ernest Rutherford's New Billiards Game', *Manchester Guardian*, 24 February 1925, 12.

32 'Breaking Up an Atom. Sir Ernest Rutherford's Experiments. A New Source of Energy', *Manchester Guardian*, 10 February 1922, 6; 'Cheap Energy from the Atom. A Doubtful Theory. Sir E. Rutherford's Views. The Laboratories' Impotence', *Manchester Guardian*, 13 September 1923, 7; 'Nature's Workshop. Electrical Atoms the One Fundamental Thing', *The Observer*, 15 March 1925, 16.

> There has never been a time when the enthusiasm of the scientific workers was greater or when there was a more hopeful feeling that great advances were imminent. . . . I feel it is a great privilege to have witnessed this period, which may almost be termed the Renaissance of Physics. . . . For example, no-one could have anticipated that the general relation between the elements would prove to be of so simple a character as we now believe it to be. It is an illustration of the fact that Nature appears to work in a simple way, and that the more fundamental the problem often the simpler are the conceptions needed for its explanation.[33]

Rutherford emphasised the primacy of engagement with nature through practical laboratory work in producing these accomplishments. 'In the main,' he argued, 'the epoch under consideration has been an age of experiment, where the experimenter has been the pioneer in the attack on new problems'. Though he recognised that the previous two decades had also been 'an age of bold ideas in theory, as the Quantum Theory and the Theory of Relativity so well illustrate', he repeatedly made the point that disciplined experimentation was the path to stable knowledge and, a fortiori, to proper scientific conduct. He had, he said, 'become more and more impressed by the power of the scientific method of extending our knowledge of Nature'. And the basis of this 'scientific method' was again '[e]xperiment, directed by the disciplined imagination either of an individual, or still better, of a group of individuals of varied mental outlook', which could 'achieve results which far transcend the imagination alone of the greatest natural philosopher'.[34]

In the unusually full public articulation of his scientific philosophy presented at Liverpool – the subject of leaders on the 'Heroic Age of Physics' and 'Truth and Simplicity' in *The Times*, no less[35] – Rutherford sought to promote his own research programme into atomic structure. This he amply achieved, with widespread and sympathetic media coverage, as we have seen. He also sought to portray the reductionist programme and modern physics generally as socially and politically unifying through their international scope:

---

33 E. Rutherford, 'The Electrical Structure of Matter', *Report of the British Association for the Advancement of Science* (1923), 1–24: 23; 'The first British Association President to Speak "Broadcast"', *Illustrated London News*, 23 September 1923, 527. For wider coverage, see, for example, 'Cheap Energy from the Atom. A Doubtful Theory. Sir E. Rutherford's Views. The Address', *Manchester Guardian*, 13 September 1923, 7–8; 'Atom that Holds Us Up. Sir Ernest Rutherford on the Mighty Units that "Build Our Universe"', *Daily Mirror*, 13 September 1923, 3. *The Times* printed a large portion of Rutherford's address verbatim over four full-length columns: 'Structure of Matter. The British Association. President's Address', *The Times*, 13 September 1923, 16.

34 Rutherford, 'The Electrical Structure of Matter' (n. 33), 23.

35 'Heroic Age of Physics', 'Truth and Simplicity', *The Times*, 13 September 1923, 11.

> In these great additions to our knowledge of the structure of matter every civilised nation has taken an active part, but we may be justly proud that this country has made many fundamental contributions. With this country I must properly include our Dominions overseas, for they have not been behindhand in their contributions to this new knowledge. It is, I am sure, a matter of pride to this country that the scientific men of our Dominions have been responsible for some of the most fundamental discoveries of this epoch, particularly in radioactivity.[36]

But this was a very particular internationalist outlook, rooted in empire and imperialism rather than the transnational cosmopolitanism we more often associate with the term in the canonical historiography. His own career, from humble origins in New Zealand, through his research training in Cambridge, his first post in Montreal, then back to Britain at Manchester and Cambridge, culminating in a BAAS Presidential Address broadcast to the nation from Liverpool, now became exemplary of imperial scientific achievement.[37] Beamed out on the BBC, reductionism and imperialist internationalism embodied in the career and person of Rutherford himself now offered a powerful combined solution to the questions of post-war political, social and natural order.

Reductionism rampant? The nuclear atom in the public sphere

Not least because of the promotional efforts of its proponents, the high profile of events like Marie Curie's visit to America in 1921 and the continuing public fascination with radium, it was hardly surprising that the atom and its spokesmen secured for themselves 'a distinctly "good press"'.[38] In this largely favourable context of reception, Rutherford's authority allowed him to exercise some control over the question of the atom as a potential source of energy – a source of continuing fascination and speculation in the media. After he had damped down such discussion in his 1923 BAAS Presidential Address, one commentator was even moved to verse on 'gathering, from the proceedings of the British Association, that [the atom] is not quite such an energetic creature as was once supposed':

---

36 Rutherford, 'The Electrical Structure of Matter' (n. 33), 22.

37 On Rutherford and empire, see L. Badash, 'The Influence of New Zealand on Rutherford's Scientific Development', in N. Reingold and M. Rothenberg (eds), *Scientific Colonialism: A Cross-Cultural Comparison* (Washington DC, 1987), 379–89; J. Campbell, *Rutherford: Scientist Supreme* (Christchurch, New Zealand, 1999).

38 J.A. Crowther, 'The Structure of the Atom', *Nature* 112 (1923), 232–3: 232.

O Atom (I write as a lover
Of restful and peaceable folk),
I own I'm relieved to discover
You're rather an indolent bloke
No longer in fierce effervescence,
I gather your energies stay
Maintained in a state of quiescence;
And I think they are better that way.

I view them, I own, with less terror
Than if they were harnessed, or hurled
(By some tiresome enthusiast's error)
To bust up the whole of our world.
Oh, vastly more noble I term it
To have strength like your own within call –
And then just to sit down like a hermit
And calmly do nothing at all.

By all means continue to coddle
Your energies safely below,
And I think you may serve as a model
For one or two people I know;
Henceforth I shall say to the bustler,
The 'live wire' and similar guys,
'Oh, go to the atom, thou hustler –
Consider his ways and be wise.'[39]

More prosaically, dozens of other commentators produced textbooks, primers and popular accounts of the emerging science of atoms and nuclei, prominent among them being Bertrand Russell's *ABC of Atoms* in 1923.[40] A particularly industrious contributor to this post-war popular physics boom was Edward Andrade, a former Manchester student of Rutherford's, whose 1923 account of *The Structure of the Atom* became a standard textbook on the new physics which went into three editions.[41] Andrade was a keen bibliophile and historian of early modern science, and the book is remarkable for its elaborate and

---

[39] Lucio, 'Lines to an Atom', *Manchester Guardian*, 15 September 1923, 7.

[40] B. Russell, *The ABC of Atoms* (London, 1923, rev. edn. 1924). Among the many other examples are F.H. Loring, *Atomic Theories* (London, 1921); J.W.N. Sullivan, *Atoms and Electrons* (London, 1923); W.F. Shearcroft, *The Story of the Atom* (London, 1925); G.P. Thomson, *The Atom* (London, 1930); C.G. Darwin, *The New Conceptions of Matter* (London, 1931).

[41] E.N. da C. Andrade, *The Structure of the Atom* (London, 1923; 2nd edn. 1924; 3rd edn. 1927). On the popular physics literature in this period see M. Whitworth, 'The Clothbound Universe: Popular Physics Books, 1919–39', *Publishing History* 40 (1996), 53–82; Bowler, *Science for All* (n. 4), esp. 34–8; H. Kragh and K.H. Nielsen,

somewhat eccentric dedication to Rutherford in seventeenth-century style – 'No astronomer of these petty Suns and Planets has given us more and rare News of them than Yourself, who first taught the Virtuosi to see in the Atom a Massy Nucleus controlling by electrick Laws his distant Servants the light-heeled Electrons' and so on, for three pages.[42] Through the 1920s, Andrade lectured widely and published a series of popular books and articles extolling the nuclear atom and its practitioners, partly countering the 'misunderstandings' of the previous generation.[43] In 1924, for example, he criticised Oliver Lodge's popular book *Atoms and Rays: An Introduction to Modern Views on Atomic Structure and Radiation* for its atavistic treatment of the ether, which, said Andrade, had no place in the 'modern' physics of atoms, nuclei and quanta.[44] Andrade pitched up against Lodge again in 1927, when the popular series Benn's Sixpenny Library published both Lodge's *Modern Scientific Ideas: Especially the Idea of Discontinuity* and Andrade's *The Atom*, a more contemporary and pro-Cavendish treatment of the topic.[45]

If Lodge, the paterfamilias of British science, continued to promote his own brand of ether physics well into the 1920s, there were other well-entrenched positions which serve to remind us that the nuclear atom was not yet orthodoxy.[46] J.J. Thomson, Master of Trinity College and still a presence in the Cavendish Laboratory and on the wider stage of British physics, was heavily antagonistic to the nuclear theory into the early 1920s, and consistently sought to play down the achievements of Rutherford and his co-workers – among them his own former assistant Francis Aston, whose invention of the mass-spectrograph and ensuing discovery of the isotopes of the light elements had provided an

'Spreading the Gospel: A Popular Book on the Bohr Atom in its Historical Context', *Annals of Science* 70 (2011), 257–83.

42 Andrade, *Structure* (n. 41), vi.

43 On Rutherford's recommendation, Andrade was appointed Professor of Physics at the Artillery College, Woolwich in 1920, and Professor of Physics at University College, London, in 1928. See A. Cottrell, 'Edward Neville da Costa Andrade', *Biographical Memoirs of Fellows of the Royal Society* 18 (1972), 1–20. Among Andrade's lectures and publications in this period are 'Atoms and Electrons: Prof. Andrade on Scientific Achievement', *The Times*, 24 February 1925, 15; *The Structure of the Atom* (n. 41); *The Atom* (London, 1927); *The Mechanism of Nature* (London, 1930).

44 E.N. da C. Andrade, 'A Veteran's View of Modern Physics', *Nature* 114 (1924), 599–601; O. Lodge, *Atoms and Rays: An Introduction to Modern Views on Atomic Structure and Radiation* (London, 1924).

45 O. Lodge, *Modern Scientific Ideas: Especially the Idea of Discontinuity* (London: 1927); E.N. da C. Andrade, *The Atom* (n. 43).

46 I. Clarke, 'Negotiating Progress: Promoting 'Modern' Physics in Britain, 1900–1940' (PhD diss., University of Manchester, 2012), esp. 103–41.

important new source of evidence for experimental and theoretical work on the nuclear atom. Aston, indeed, became a significant proselyte for the nuclear theory and an advocate of a fairly aggressive reductionist conception of nuclear structure (Figure 4.1).[47] By 1923, after Aston and Soddy had won Nobel Prizes in Chemistry for their work on isotopes, and Bohr the Nobel Prize in Physics for his work on atomic structure – based, as it was, on Rutherford's nuclear model – Thomson had grudgingly come to accept the nuclear atom.[48] Now, though, he tried to turn it to his advantage in ways that helped fulfil his own long-standing ambition to unite the physical sciences through his own theory of 'the electron in chemistry'.[49]

If Thomson sought his own path to unification through reduction (a path as yet insufficiently explored in the historical literature[50]), others within the British physics community dissented from Rutherford's view both as to the self-evident significance of 'pure', 'fundamental', reductionist research, and as to the importance of imperial internationalism in the quest for scientific truth.[51] But the emerging consensus was strengthened in the public sphere in other ways. At the British Empire Exhibition at Wembley in 1924 and 1925, for example, an 'Exhibition of Pure Science' was organised under the auspices of the Royal Society. Chaired by F.E. Smith of the Department of Scientific and Industrial Research, the organising committee included most of the great and the good of British physics: Lodge, Bragg, Richard Glazebrook, Arthur Schuster, Napier Shaw, Campbell Swinton and the Director of the Science Museum, Henry Lyons. The exhibition was intended to illustrate 'recent British research in the fundamental branches of pure science', and drew heavily on the work of the

---

47 F.W. Aston, 'The Atoms of Matter: Their Size, Number, and Construction', *Nature* 110 (1922), 702–5; idem, *Isotopes* (London, 1922); G. Hevesy, 'Francis William Aston, 1877–1945', *Obituary Notices of the Fellows of the Royal Society* 5 (1948), 635–50; Hughes, 'Making Isotopes Matter' (n. 29).

48 Friedman, *Politics of Excellence* (n. 9), 184–5; Kragh, *Niels Bohr and the Quantum Atom* (n. 20), 271ff.

49 J.J. Thomson, *The Electron in Chemistry* (London, 1923); S.B. Sinclair, 'J.J. Thomson and the Chemical Atom: From Ether Vortex to Atomic Decay', *Ambix* 34 (1987), 89–116.

50 But see M. Chayut, 'J.J. Thomson: The Discovery of the Electron and the Chemists', *Annals of Science* 48 (1991), 527–44; K. Gavroglu, 'The Physicists' Electron and its Appropriation by the Chemists', in Buchwald and Warwick (eds), *Histories of the Electron* (n. 16), 363–400.

51 See, for example, the contrasting views in 'Sir E. Rutherford on Atomic Discoveries', *The Times*, 11 August 1924, 10, and W.H. Bragg, 'Research Work and its Applications', *Nature* 113 (1924), 311–12. This point is explored at greater length in J. Hughes, 'Redefining the Context: Oxford and the Wider World of British Physics, 1900–1940', in R. Fox and G. Gooday (eds), *Physics in Oxford, 1839–1939* (Oxford, 2005), 267–300.

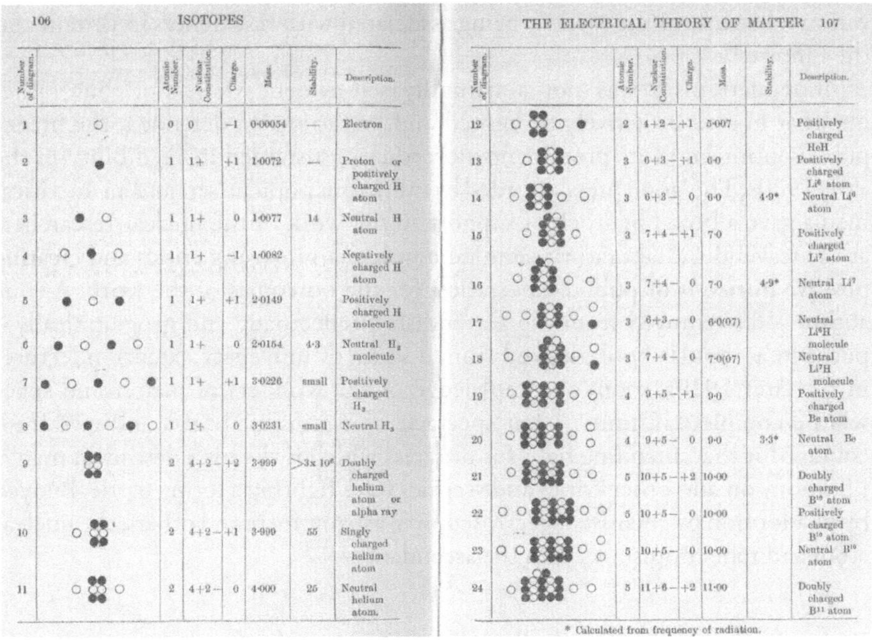
106 ISOTOPES

Number of diagram.		Atomic Number.	Nuclear Constitution.	Charge.	Mass.	Stability.	Description.
1		0	0	−1	0·00054	—	Electron
2		1	1+	+1	1·0072	—	Proton or positively charged H atom
3		1	1+	0	1·0077	14	Neutral H atom
4		1	1+	−1	1·0082		Negatively charged H atom
5		1	1+	+1	2·0149		Positively charged H molecule
6		1	1+	0	2·0154	4·3	Neutral H_2 molecule
7		1	1+	+1	3·0226	small	Positively charged H_3
8		1	1+	0	3·0231	small	Neutral H_3
9		2	4+2−	+2	3·999	$>3\times10^6$	Doubly charged helium atom or alpha ray
10		2	4+2−	+1	3·999	55	Singly charged helium atom
11		2	4+2−	0	4·000	25	Neutral helium atom.

THE ELECTRICAL THEORY OF MATTER 107

Number of diagram.		Atomic Number.	Nuclear Constitution.	Charge.	Mass.	Stability.	Description.
12		2	4+2−	+1	5·007		Positively charged HeH
13		3	6+3−	+1	6·0		Positively charged Li^6 atom
14		3	6+3−	0	6·0	4·9*	Neutral Li^6 atom
15		3	7+4−	+1	7·0		Positively charged Li^7 atom
16		3	7+4−	0	7·0	4·9*	Neutral Li^7 atom
17		3	6+3−	0	6·0(07)		Neutral Li^6H molecule
18		3	7+4−	0	7·0(07)		Neutral Li^7H molecule
19		4	9+5−	+1	9·0		Positively charged Be atom
20		4	9+5−	0	9·0	3·3*	Neutral Be atom
21		5	10+5−	+2	10·00		Doubly charged B^{10} atom
22		5	10+5−	0	10·00		Positively charged B^{10} atom
23		5	10+5−	0	10·00		Neutral B^{10} atom
24		5	11+6−	+2	11·00		Doubly charged B^{11} atom

* Calculated from frequency of radiation.

Figure 4.1 Aston's scheme for the building up of isotopic nuclei from protons and electrons.

Source: F.W. Aston, *Isotopes* (London, 1922).

Cavendish Laboratory, whose contribution was coordinated by Patrick Blackett, one of Rutherford's up-and-coming protégés.[52] The exhibition began with the 'discovery of the electron' in 1897 (an achievement now unproblematically attributed to J.J. Thomson, despite its earlier contested status), and used the Rutherford-Bohr atomic theory to order the exhibits (including C.T.R. Wilson's cloud chamber and F.W. Aston's first mass spectrograph) in a linear historical sequence, 'a story of discovery and progress in both knowledge and technique'.[53] The exhibition served a three-fold function. In addition to promoting the reductionist regime in a major show attended by some 17.5 million visitors, it strongly linked this work to Cambridge and the Cavendish. Significantly, it also endorsed a nationalist view of the history of subatomic physics by portraying

52 M.J. Nye, *Blackett: Physics, War and Politics in the Twentieth Century* (Cambridge, MA, 2004).

53 A.Q. Morton, 'The Electron Made Public: The Exhibition of Pure Science in the British Empire Exhibition, 1924–5', in B. Finn (ed.), *Exposing Electronics* (London, 2003), 25–43; D.R. Knight and A.D. Sabey, *The Lion Roars at Wembley: The British Empire Exhibition 1924–1925* (London, 1984).

various elements of the work as being associated with researchers in Britain and the Empire.[54]

If reductionism was not a naturally self-evident viewpoint, but rather one that had to be actively promoted and, occasionally, defended, the breezy public optimism of its proponents nevertheless sustained its credibility in the early 1920s. The 'good press' afforded by numerous popularisers and an uncritical media gave a broad bedrock of support to the work of the nuclear researchers at the Cavendish Laboratory, simultaneously marginalising critics and creating positive horizons of public expectation for the outcomes of the work. As one author – far removed from the Cavendish intellectually and geographically – put it in a popular book derived from a series of university extension lectures in the later 1920s, soon 'our knowledge of the wonders of matter and space will be complete in unity, a fair spectacle to warm our hearts and give fresh courage for that unending fight for progress which is the high destiny of man'.[55] J.J. Thomson and other critics and sceptics were fighting a losing battle. Buoyed by its adoption by chemists, spectroscopists, astronomers and others, the nuclear atom and reductionism were in the ascendant.

Reductionism undermined: discipline, dissent and reform

That the nuclear atom and the reductionist programme attracted considerable public attention and commanded intellectual authority in the early 1920s was possible in large measure because Rutherford and the Cavendish largely had the field of nuclear research to themselves. From 1923, however, another laboratory became involved in nuclear disintegration experiments. Deploying ostensibly the 'same' experimental methods as those used in Cambridge, a group of researchers led by Hans Pettersson and Gerhard Kirsch at the Institut für Radiumforschung in Vienna, systematically repeated and extended the Cambridge experiments. Their work suggested that many more elements could be disintegrated than had been found to be the case in Cambridge. Because Rutherford's theorising as to the structure of nuclei was based on the experimental results achieved in his own laboratory, the Viennese work challenged not only the experimental 'facts' but all the deductions based upon them, and therefore Rutherford's carefully

---

54 Royal Society, *British Empire Exhibition 1924: Handbook to the Exhibition of Pure Science* (London, 1924); Royal Society, *Phases of Modern Science: Published in Connexion with the Science Exhibit Arranged by a Committee of the Royal Society in the Pavilion of His Majesty's Government at the British Empire Exhibition, 1925* (London, 1925); Clarke, 'Negotiating Progress', (n. 46), 175–83.

55 R.G. Lunnon, *New Worlds for Old: The Realm of Modern Physics* (London: 1928), 99. Lunnon was Lecturer in Physics at Armstrong College, University of Durham.

reasoned account of nuclear constitution.[56] Over the next five years, the Vienna workers doggedly pursued their work and their opposition to the Cambridge results, developing state-of-the-art techniques of nuclear research in order to do so (Figure 4.2). And, just as Rutherford and his allies had done, Pettersson worked hard to popularise and publicise his work to wide audiences.[57]

In maintaining a firm distinction between the private and epistemically contingent workspace of the laboratory and the public and epistemically constitutive display spaces of the lecture room and the media, what Rutherford had not revealed in his flag-waving speeches at the RI, the BAAS and elsewhere was the *difficulty* and *doubt* surrounding the experiments carried out in his laboratory.[58] The results of nuclear disintegration experiments were detected using the scintillation technique, in which the particles resulting from an artificially induced nuclear disintegration created a momentary flash when they impacted on a small zinc sulphide screen. In a darkened room, these flashes could be observed through a telescope, counted, and the results used to deduce what had taken place during the experiment – for example, whether a particular target nucleus had undergone disintegration. This was the technique used in the experiments which had revealed the nuclear atom over a decade earlier, and which had been the basis of all nuclear disintegration experiments (and all the theorising as to nuclear structure based on them) since. But counting scintillations was a tiring and difficult task, fraught with pitfalls and constantly liable to be undermined by radioactive contamination or human error. Elaborate protocols were required in order to ensure consistency between one count and the next, between different runs of the same experiment and between different series of experiments. Within the Cavendish, these protocols were

---

56 R.H. Stuewer, 'Artificial Disintegration and the Cambridge-Vienna Controversy', in P. Achinstein and O. Hannaway (eds), *Observation, Experiment and Hypothesis in Modern Physical Science* (Cambridge, MA, 1985), 239–307. For a revisionist account of the controversy and its consequences, see Hughes, 'The Radioactivists' (n. 14), 94–120; J. Hughes, 'The French Connection: The Joliot-Curies and Nuclear Research in Paris, 1925–1933', *History and Technology* 13 (1997), 325–43; idem, '"Modernists with a Vengeance": Changing Cultures of Theory in Nuclear Science, 1920–1930', *Studies in History and Philosophy of Modern Physics* 29 (1998), 339–67; idem, 'Plasticine and Valves' (n. 29). A richer analysis of the Vienna position is given in Rentetzi, *Trafficking Materials* (n. 13), 137–80.

57 H. Pettersson, *Atomernas Sprängning. En Studie i Modern Alkemi* (Stockholm, 1927).

58 On the public/private distinction, see: D. Gooding, 'In Nature's School: Faraday as an Experimentalist', in D. Gooding and F.A.J.L. James (eds), *Faraday Rediscovered: Essays on the Life and Work of Michael Faraday* (London, 1985), 105–35. On the epistemologically contingent versus constitutive, see: G.N. Gilbert and M. Mulkay, *Opening Pandora's Box: A Sociological Analysis of Scientists' Discourse* (Cambridge, 1984). On the subject more generally: H. Collins, 'Public Experiments and Displays of Virtuosity: The Core-Set Revisited', *Social Studies of Science* 18 (1988), 725–48.

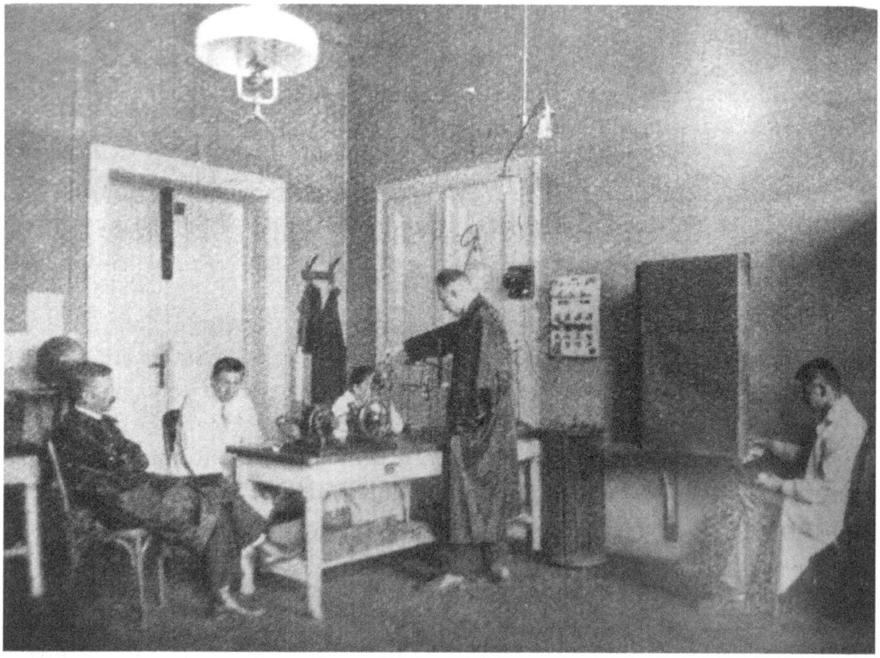

Figure 4.2 Hans Pettersson and his co-workers at the Institut für Radiumforschung, Vienna.

Source: H. Pettersson, *Atomernas Sprängning. En Studie i Modern Alkemi* (Stockholm, 1927). Thanks to Petter Hellström for helping to retrieve the picture.

managed by Rutherford's deputy, James Chadwick. And within the confines of the one laboratory, they could be made to work reasonably well (though the laboratory notebooks were still riddled with comments indicating doubt about the certitude of the results of many series of scintillation counts). But when a second laboratory entered the field and did not necessarily observe the same protocols, the difficulties of replication became all too apparent.[59]

Protracted private correspondence between the various parties did little to resolve the issues under dispute, and there were even *sotto voce* imputations of improper conduct.[60] An indication of the way the battle lines of the controversy were drawn is given by the tone of a private remark by Chadwick to Rutherford:

59 Hughes, 'The Radioactivists' (n. 14), 94–120. The classic study of the problems of replication and their epistemological consequences is H.M. Collins, *Changing Order: Replication and Induction in Scientific Practice* (London, 1985).

60 Stuewer, 'Artificial Disintegration' (n. 56), 248–9.

> Our friend Kirsch has now let himself loose in the Physikalische Zeitschrift. His tone is really impudent, to put it very mildly. . . . Kirsch & Pettersson seem to be rather above themselves. A good kick from behind would do them a lot of good. The name on the paper is that of Kirsch but the voice is the familiar bleat of Pettersson. I don't know which is the boss but as Mr Johnson said there is no settling a point of precedence between a louse and a flea.[61]

The controversy dragged on, with neither side able to establish an independent basis for establishing the 'correct' outcome to the experiments and both claiming the legitimacy of their own methods and interpretations. By 1927 the two groups had reached an impasse – a classic example of the 'experimenters' regress'.[62]

The Cambridge-Vienna situation was exacerbated by a parallel controversy between Cambridge's Charles Ellis and Lise Meitner in Berlin concerning the nature and interpretation of the beta-ray spectrum, in which issues of experimental technique and theoretical interpretation were again at stake, and in which there were also significant implications for the validity of Rutherford's nuclear model.[63] The stakes were therefore high. As we have seen, the Cavendish had established an authoritative public position for its work, and Rutherford's own professional credibility and prestige hinged on the success of the lab's flagship research programme. Even more significantly, the legitimacy of the whole reductionist regime was now potentially under threat, because the controversy with the Viennese threatened to expose the carefully concealed contingency of the Cambridge results and thereby emphasise the fragility of the experimentally grounded reductionist programme. After all, Rutherford's public pronouncements about the epistemic primacy of experiment were not simply generalised assertions about the scientific method, but were also claims for his reductionist physics to be *exemplary* of it. Yet with journal publications of claim and counterclaim, the two controversies could not but attract increasing attention from the wider scientific public and beyond.

Rutherford's own professional status added to the urgency of a rapid and carefully managed resolution to the controversies besetting his lab and the reductionist project. He became President of the Royal Society in late 1925, and his position as the official spokesman for British science can only have served to exacerbate the embarrassment caused by his involvement in the increasingly acrimonious controversy with Kirsch and Pettersson. In the changing international context, Rutherford and the Royal Society took a leading role in restoring the Central Powers to membership of the International Research

---

61 Chadwick to Rutherford, [July–August 1925], quoted in ibid, 267.

62 H.M. Collins, *Changing Order* (n. 59).

63 C. Jensen, *Controversy and Consensus: Nuclear Beta Decay 1911–1934* (Basel, 2000).

Council in 1926.[64] Rutherford now used the rhetoric of internationalism, rather than imperialism, noting in his first presidential address to the Royal Society in 1926 that for two successive years the Society's council 'had unanimously supported the admission of the Central Powers to full membership of the International Research Council'. He recorded his certitude that 'all members of the Society will welcome the disappearance of restrictions in membership which were seriously hampering the international co-operation of scientific men', and that they might now 'look forward hopefully to an ever-growing friendship and co-operation between the scientific men of different nations, thus promoting not only the advance of science but goodwill between the nations'.[65] Overt controversy between his own lab and others in Berlin and Vienna would hardly have helped that case.

The solution to this problem of credibility and authority, both within the discipline and on a broader stage, was active private management of an increasingly public controversy. Within the laboratory, as Chadwick's comments to Rutherford indicate, the dispute often followed a deeply antagonistic course. Yet in correspondence with people outside the Cavendish, Rutherford took a much more irenic approach, seeking to contain the damaging disagreements within the private sphere of personal communication characteristic of the long-established radioactivity network. He wrote in confidence to Pettersson's boss Stefan Meyer and to Niels Bohr (both long-standing friends and confidants) to warn them of the Swede's work: '[Pettersson] seems a clever and ingenious fellow, but with a terrible capacity for getting hold of the wrong end of the stick. From our experiments Chadwick and I are convinced that nearly all his work . . . is either demonstrably wrong or wrongly interpreted.'[66] He now insisted on the containment of dissent and on the *private* conduct of the controversy as part of a strategy to maintain the public fiction he had helped create in his heroic speeches from numerous public platforms over the previous few years. As he had put it to Pettersson, 'it is better to discuss these divergences of view in private than in print. Workers in this field are too few and too select to misunderstand one another'.[67] Thus Rutherford now appealed not

---

64 On the IRC, see A.G. Cock, 'Chauvinism and Internationalism in Science: The International Research Council, 1919–1926', *Notes and Records of the Royal Society* 37 (1983), 249–88; F. Greenaway, *Science International: A History of the International Council of Scientific Unions* (Cambridge, 1996), 19–32.

65 'Address of the President, Sir Ernest Rutherford, at the Anniversary Meeting, 30 November 1926', *Proceedings of the Royal Society* A, 113 (1927), 481–95: 486.

66 Rutherford to Bohr, 18 July 1924, quoted in Stuewer, 'Artificial Disintegration' (n. 56), 255.

67 Rutherford to Pettersson, 19 July 1924, quoted in Stuewer, 'Artificial Disintegration' (n. 56), 256.

to nature or to scientific method as the proper grounds for scientific concord, but to professional and disciplinary solidarity. Scientific unity was to be maintained and guaranteed not by nature but by social unity. As Shapin and Schaffer put it, '[s]olutions to the problem of knowledge are solutions to the problem of social order'.[68]

While the beta-ray controversy was ultimately resolved amicably, the Cambridge-Vienna controversy came to a head in December 1927, when Chadwick himself visited Vienna. By running a series of control experiments, he was able to demonstrate crucial differences between the protocols of the Vienna and the Cambridge scintillation-counting experiments. The Viennese counters – a team of young women – seemed to know the expected outcome of experiments in their work with Pettersson. When Chadwick took them through a series of 'blind' experimental runs, the results resembled more closely those found in Cambridge, thus apparently vindicating the Cavendish position. In keeping with Rutherford's policy of containment rather than confrontation in controversy, Chadwick and Meyer agreed to keep this denouement private so as to avoid further public discussion. Following Chadwick's visit, Rutherford wrote to Pettersson:

> I hope that the conversations with Chadwick will be helpful to a better understanding amongst us all. There are so few workers in this difficult subject that we must try and pull together and settle our differences as far as possible by private correspondence rather than by controversies in scientific journals, which in my experience do nothing but cause irritation. If you and your friends are of the same opinion I think there should be no great difficulty in settling our differences.[69]

This was an attempt to ensure that damaging details of the contingencies of the scintillation counting experiments were not unnecessarily exposed to the public gaze. In fact, the Viennese continued to argue their case – not, now, on the basis of the scintillation-counting experiments, but through other instruments, including cloud chambers and a mass spectrograph.[70] Yet within months of Chadwick's visit to Vienna, new developments in instrumentation made the scintillation-counting technique, with all its difficulties and doubts, obsolescent. The Cambridge-Vienna controversy did not achieve a definitive closure: it quickly became irrelevant, as one kind of unity was displaced by another.

---

68 S. Shapin and S. Schaffer, *Leviathan and the Air Pump: Hobbes, Boyle and the Experimental Life* (Princeton, 1985), 332.

69 Rutherford to Pettersson, 9 January 1928, quoted in Stuewer, 'Artificial Disintegration' (n. 56), 289.

70 Rentetzi, *Trafficking Materials* (n.13), 180–226.

Material culture, the media and the rise of nuclear physics, 1928–34

The impact of the Cambridge-Vienna controversy on Rutherford's public discourse was profound. From 1924 to the early 1930s, there were no bullish lectures on reductionism or the exemplary nature of nuclear science. In his presidential addresses to the Royal Society, for example, Rutherford for the most part took pains not to speak about his own flagship nuclear physics, as one might have expected, but about wireless research and the upper atmosphere (1926) and about recent work on intense magnetic fields and high voltages (1927) – both topics also associated with the Cavendish Laboratory, but not with Rutherford personally. In response to the controversies of the 1920s, the Cambridge group organised a conference in the summer of 1928 to discuss problems in radioactivity. Most of the key European and American workers in the subject were invited (though not those from Vienna), and as a result of the discussions at the meeting, several reoriented their research towards the contested artificial disintegration experiments. They drew on a new element of technique: electronic particle counters. In 1928 Geiger and Müller unveiled an electrical counter capable of counting particles under a variety of conditions.[71] With the contemporaneous development of relatively reliable and inexpensive electronic components in the booming radio industry and the newly available and widely distributed skills of young wireless enthusiasts in the late 1920s, it became possible to construct stable – and *reproducible* – electronic counting equipment for use in the laboratory. At the Cavendish, dedicated wireless enthusiasts like Eryl Wynn-Williams and Bennett Lewis drew on the standardised valves and other products of the radio industry, appropriately adapted to the laboratory setting, to make and calibrate reliable amplifiers and complex counting circuits.[72] Aided by circuit diagram blueprints freely circulated to all who asked for them and the widely distributed skills of radio hams in physics departments elsewhere, the number of sites at which electronic detectors were developed grew rapidly so that by 1930 electronic particle counters were in regular use in disintegration experiments in several laboratories, including those in Cambridge and Vienna, Berlin, Paris, Halle, Kiel and Giessen, with many others following in later years.[73]

This expansion of the number of laboratories engaged in active work on the disintegration experiments fundamentally reshaped nuclear research around a new set of tools and a new set of questions. The shift was reinforced by a simultaneous set of changes in the relationship between experimentalists and

---

[71] T.J. Trenn, 'The Geiger-Müller Counter of 1928', *Annals of Science* 43 (1986), 111–35.

[72] This account is developed in more detail in Hughes, 'Plasticine and Valves', (n. 29).

[73] On the broader significance of amateur wireless in this period, see K. Haring, *Ham Radio's Technical Culture* (Cambridge, MA, 2007).

mathematical theoreticians. The 1920s saw significant activity in theoretical atomic physics at Bohr's Institute for Theoretical Physics in Copenhagen and elsewhere. With the support of the Rockefeller Foundation and other philanthropic bodies a new generation of students turned their attention to atomic theory, forming a small and highly mobile international community. The development of wave mechanics from 1926, and its application to the nucleus from 1928 (after, and arguably in *response* to, the crisis of certitude prompted by the Cambridge-Vienna controversy), offered a new resource for the understanding of nuclear phenomena. In particular, the development of the notions of nuclear energy levels, quantum tunnelling and of resonance nuclear penetration allowed experimentalists to focus on novel kinds of phenomena using the new electrical techniques now increasingly at their disposal. Experimentalists' adoption of wave mechanics simultaneously legitimated the new mathematics and its practitioners and restructured the relationship between laboratory researchers and the growing community of mathematical theorists, creating a new and mutually reinforcing dialogue – hence theorist George Gamow's dedication of his 1931 book *The Constitution of Atomic Nuclei and Radioactivity* to the Cavendish Laboratory.[74]

Out of this new confluence of experimental and theoretical technique flowed new modes of practice, new epistemological commitments and new discoveries. The neutron, the deuteron, the positron and other discoveries in the early 1930s offered productive new foci for research, though again controversy and interlaboratory rivalry, rather than conceptual consensus or methodological unity, were the drivers for advance.[75] And the most visible aspect of the changes in nuclear technique of the early 1930s, the development of particle accelerators, initiated a race to ever-higher machine energies and a research regime which persists into the twenty-first century at CERN and elsewhere.[76] Importantly, however, all these new discoveries were produced and sustained by a range of novel and widely shared experimental techniques, and they mapped the growth of a new subdiscipline – nuclear physics – and a new scientific identity: the nuclear physicist.

---

74 For further detail see Hughes, '"Modernists with a Vengeance"' (n. 56). G. Gamow, *The Constitution of Atomic Nuclei and Radioactivity*, (Oxford, 1931); Stuewer, 'Gamow's Theory of Alpha Decay' (n. 29).

75 R.H. Stuewer, 'The Naming of the Deuteron', *American Journal of Physics* 54(3) (1986), 206–18; idem, 'Mass-Energy and the Neutron in the Early Thirties', *Science in Context* 6 (1993), 195–238; idem, The Origin of the Liquid-Drop Model and the Interpretation of Nuclear Fission', *Perspectives on Science* 2 (1994), 76–129; X. Roqué, 'The Manufacture of the Positron', *Studies in History and Philosophy of Modern Physics* 28 (1997), 73–129.

76 J.L. Heilbron and R.W. Seidel, *Lawrence and his Laboratory: A History of the Lawrence Berkeley Laboratory* (Berkeley, 1989); Hughes, *The Manhattan Project* (n. 5).

The new discoveries of the early 1930s were not a vindication of the reductionist programme, but a consequence of a new material and sociotechnical culture. Epistemological unification – insofar as there was unification, given the continuing controversies in nuclear physics in the 1930s – depended on the material unification provided by particular kinds of electronic valves, plasticine and other commercially available products. By this time, in any case, the changing economic and political context was impacting on public scientific discourse, especially perhaps that of 'pure' science and reductionist unificationism. In the later 1920s, Rutherford's swaggering attitude to science changed as he became a powerful man of affairs with broader responsibility for scientific administration – not only as President of the Royal Society (1925–30), but also in his work with the BBC, including service on its General Advisory Council, and as Chairman of the Science Advisory Council of the Department of Scientific and Industrial Research (DSIR, 1930–37). In these posts, he had of necessity to take a broader view of science in the *national* setting, particularly in the wake of the Depression, when funding was stretched and science was often scapegoated as the cause of economic ills.[77] Science in industry became the catchphrase for public scientific discourse, and by 1934, Rutherford was explicitly expressing worries about the 'startling rapidity' of social and economic change, largely brought on by the advances in science and technology that he amongst others had done so much to promote over previous years.[78] He worried, too, about criticism from members of the DSIR Advisory Council and others about the uselessness to industrial requirements of nuclear research and the PhDs he was churning out at Cambridge.[79]

But with the consolidation of extreme forms of nationalism in Europe and the reorientation of national scientific agendas in the early 1930s, the new prestige of nuclear physics in its modern, high-tech guise offered considerable rewards for some. The entry of Enrico Fermi's Rome group of physicists into the nuclear field in the early 1930s, for example, speaks more to a nationalist agenda to enter a high-prestige scientific field with large potential intellectual rewards than to any concerns of internationalism, the self-evident significance

---

77 Wilson, *Rutherford* (n. 11), 453–95; C. Pursell, '"A Savage Struck by Lightning": The Idea of a Research Moratorium, 1927–37', *Lex et scientia* 10 (1974), 146–58; R. MacLeod and K. MacLeod, 'The Social Relations of Science and Technology, 1914–1939', in C.M. Cipolla (ed.), *The Fontana Economic History of Europe: The Twentieth Century – 1* (Glasgow, 1976), 301–63; A.-K. Mayer, '"A combative sense of duty": Englishness and the scientists', in Lawrence and Mayer, (eds), *Regenerating England* (n. 22), 67–106.

78 See, for example, the comments in E. Rutherford, 'Trends in Modern Physics', *Proceedings of the Society of Chemical Industry* 16 (1934), 69–71: 69. This context is amplified in J. Hughes, '"Divine Right" or Democracy? The Royal Society "Revolt" of 1935', *Notes and Records of the Royal Society* 64 (2010), S101–S117, esp. S104–S106.

79 M. Oliphant, *Rutherford: Recollections of the Cambridge Days* (Amsterdam, 1972), 145–6.

of the reductionist programme or even its prospective utility.[80] As Rutherford worked with George Trevelyan, William Beveridge and Frederick Gowland Hopkins in 1933 to set up the Academic Assistance Council, which took a lead in aiding displaced Jewish academic refugees from fascist European countries, internationalism again became a powerful trope, though it was now in tension with the increasingly dominant discourse of science and the national economy.[81] Here, even reductionist science now served a very different purpose than it had a decade earlier. Far from being disconnected from social concerns, it now had to be seen to play its part in the creation, not the elimination, of jobs for national economic benefit. As political conditions changed, a movement that had been intended to unify science now became part of a movement of *national* unity. In his 1936 Norman Lockyer Lecture to the British Science Guild, Rutherford explicitly linked the development of 'pure' science to the development of laboratory technique – principally equipment for the production of high vacua – and, in an early statement of what would come to be known as the 'linear model' of innovation, drew attention to the interplay between academic research and industrial research and production and the 'orderly application of scientific research to industrial ends'.[82] The necessity for his argument that '[t]hese remarkable developments have not led to any serious displacement of labour in older industries, but have rather given new avenues of employment for great numbers of men and women' – about 16,000 in the British lamp and valve industry alone, for example – would have been unthinkable in the optimism surrounding reductionist science a decade earlier.[83]

---

80 G. Holton, 'Striking Gold: Fermi's Group and the Recapture of Italy's Place in Physics', *Minerva* 12 (1974), 159–98; L. Orlando, 'Physics in the 1930s: Jewish Physicists' Contribution to the Realisation of the "New Tasks" of Physics in Italy', *Historical Studies in the Physical and Biological Sciences* 29 (1998), 141–81; G. Battimelli, 'Funds and Failures: The Political Economy of Fermi's Group', in *Proceedings of the International Conference 'Enrico Fermi and the Universe of Physics', Rome, September 29–October 2, 2001* (Rome, 2003), 169–84.

81 Wilson, *Rutherford* (n. 11), 483–89; D. Zimmerman, 'The Society for the Protection of Science and Learning and the Politicization of British science in the 1930s', *Minerva* 44 (2006), 25–45; N. Cameron, 'The Politics of British Science in the Munich Era', in Shea (ed.), *Otto Hahn* (n. 29), 181–99.

82 Rutherford, *Science in Development* (London, 1936), 17.

83 On the changing connotations of 'pure', 'fundamental' and 'applied' science in this period, see S. Clarke, 'Pure Science with a Practical Aim: The Meanings of Fundamental Research in Britain, circa 1916–1950', *Isis* 101 (2010), 285–311; R. Bud, '"Applied Science": A Phrase in Search of a Meaning', *Isis* 103 (2012), 537–45; G. Gooday, '"Vague and Artificial": The Historically Elusive Distinction Between Pure and Applied Science', *Isis* 103 (2012), 546–54.

In a changing political and economic context, in which 'pure' science and the Cavendish experimental programme and its reductionist underpinnings had to be defended and justified in ways carefully attuned to national values and goals, the fortuitous and redemptive discoveries of the early 1930s were mediated to the public by a new breed of science correspondents who took their values more or less directly from the nuclear scientists themselves. Leading the pack was J.G. Crowther, science correspondent of the *Manchester Guardian*, a Marxist and a prolific commentator on 1930s scientific developments. Crowther had become acquainted with the Cavendish scientists in the mid-1920s in his role as a technical representative for Oxford University Press. He had forged friendships with Kapitza, the Russian physicist working on the development of intense electric and magnetic fields, and with John Cockcroft, a close colleague of Kapitza's, one of Rutherford's lieutenants in the management of the Cavendish in the 1930s, and one of Crowther's companions on a visit to the Soviet Union in 1931.[84] In the later 1920s, Crowther began writing occasional articles on aspects of science for periodicals such as the *New Statesman* and newspapers such as the *Manchester Guardian*, some of which were collected in books like *Science for You* (1928) and *Osiris and the Atom* (1932).[85] Crowther quickly became an admirer of the Cavendish – indeed, one of the chapters in *Science for You* paid breathless homage to the 'brilliance' and 'eminence' of the lab and the 'unparalleled sequence of genius' constituted by its directors.[86] He evidently came quickly to Rutherford's notice, for by 1929 Rutherford was writing him references for job and grant applications. In 1931, Crowther (working with Cockcroft) played an important role in coordinating press coverage of the history of science extravaganza built around the centenary of Faraday's discovery of electromagnetism and the centenary of the birth of James Clerk Maxwell, the first director of the Cavendish.[87] Through this he became a favoured press contact of the Cavendish scientists.

It was to Crowther that the Cavendish turned again in early 1932 to manage media coverage of emerging new lines of research. Crowther was invited to a session of the Kapitza Club – an informal discussion group organised by the Russian physicist to discuss recent research – on Chadwick's discovery of the neutron in February 1932. His accounts in the *Manchester Guardian*, widely

---

84 J.W. Boag, P.E. Rubinin and D. Shoenberg (eds), *Kapitza in Cambridge and Moscow: Life and Letters of a Russian Physicist* (Amsterdam, 1990); G. Hartcup and T. Allibone, *Cockcroft and the Atom* (Bristol, 1984).

85 J.G. Crowther, *Science for You* (London, 1928); idem, *Osiris and the Atom* (London, 1932); idem, *Fifty Years with Science* (London, 1970).

86 Crowther, *Science for You* (n. 85), 112.

87 F.A.J.L James, 'Presidential Address: The Janus Face of Modernity: Michael Faraday in the Twentieth Century', *British Journal for the History of Science* 41 (2008), 477–516.

Figure 4.3 John Cockcroft with the Cavendish apparatus for the artificial acceleration of protons, as illustrated in Crowther's book *The Progress of Science* (London, 1934).

disseminated in other newspapers, deliberately laid the discovery out as a triumph of international and reductionist science, the Cavendish providing the final and inspirational climax to a sequence of work carried out in Berlin and Paris. A few weeks after the neutron scoop, to Cockcroft's fury, Crowther was attending a conference at Bohr's Institute in Copenhagen when research by Cockcroft and Walton on the use of artificially accelerated protons to disintegrate nuclei came to fruition. On his return, Crowther was quickly lined up to coordinate press coverage. He told his editor, 'I now find that I am becoming as it were the press-agent of the Cavendish Laboratory', and disclosed that early hints and exclusive stories were to be given to him in return for his efforts on the laboratory's behalf.[88] While there were still occasional tensions between Crowther and the *Guardian*'s editorial staff, he went on to work closely with Cockcroft and others to cover the work of the Cavendish in highly favourable terms (Figure 4.3).[89] Crowther's promotion of the emergent and – as we have seen, still heavily contested – field of nuclear physics as an internationalist, epistemologically transparent and ontologically privileged enterprise did much to entrench the discipline and the reductionist programme in the public sphere at a politically and economically difficult moment for the Cavendish.[90] What is perhaps most striking is that in uncritically reproducing the values of the physicists themselves, Crowther seems to have put the Cavendish programme and the reductionist project outside his own, Marxist, frame of political analysis.[91] Here was an unexpected form of unity, indeed.

Conclusion

If C.P. Snow did not win a Cambridge studentship, not least because he was up against one of 'Rutherford's men', by 1933 he had nevertheless become an ardent admirer of the reductionist project. In an overview of 'Cambridge chemistry', he began with an account of atomic structure, because 'atoms are the bricks from which molecules are built', and 'this physical knowledge is the starting-point for chemistry, which is concerned with the molecules when formed'. Castigating much of traditional chemistry as mere 'recipes', Snow insisted that the 'modern' subject 'deals with the structure of molecules and their reactions', and that the 'more detail that can be obtained upon the changes in the outer (or "binding")

---

88 J.G. Crowther to W.P. Crozier, 9 May 1932, quoted in J. Hughes, 'Nuclear Physics at the Cavendish Laboratory in the Thirties', in *Proceedings* (n. 80), 105–17: 111. Crowther claimed similar privileged access to the Cambridge biochemists, on whom see M. Weatherall and H. Kamminga, *Dynamic Science: Biochemistry in Cambridge 1898–1949* (Cambridge, 1992).

89 These tensions are explored in J. Hughes, 'Insects or Neutrons? Science News Values in Interwar Britain', in M.W. Bauer and M. Bucchi (eds) *Journalism, Science and Society: Science Communication Between News and Public Relations* (New York, 2007), 11–20.

90 See, for example, his treatment of the Cavendish in J.G. Crowther, *The Progress of Science: An Account of Recent Fundamental Researches in Physics, Chemistry and Biology* (London, 1934), 1–53.

91 C. Chilvers, 'The Dilemmas of Seditious Men: The Crowther-Hessen Correspondence in the 1930s', *British Journal for the History of Science* 36 (2003), 417–35.

electrons when molecules react, the more rigid and satisfying the science will become'.[92] At a broader level, too, Snow emphasised the significance of the reductionist project. Science, he asserted, was a collective endeavour. Most scientists were merely 'filling in gaps in knowledge; they are doing nothing which could be called by any stretch of the imagination "original"'. 'The individual matters very little', he suggested, and it was 'only the scientist of genius who can add greatly to the rate of progress'. The examples of such genius he chose were not chemists: 'Rutherford and Bohr are possibly the only scientists of the twentieth century who have anticipated the normal progress of their time by more than twenty years'.[93]

Rutherford was, for Snow, clearly one of the 'two Englishmen living to-day whom no one, however deterred by caution or jealousy or all the other motives which lead one to extol the past at the expense of the present, could avoid putting in the highest class of all, the class which contains Newton and Faraday, and almost no one else'.[94] The rest he damned with faint praise: 'By far the greater proportion of scientists are competent workers doing competent work in a restricted section of a field which is already mapped' – 'children of their time benefiting by the unusual good luck of being alive during some splendidly rich years'. His own subject, molecular spectroscopy, he now conceded, was 'like unravelling an immense and rather tedious crossword puzzle', and '[a]nyone who has solved a Torquemada puzzle would be able, with a little training, to become a successful molecular spectroscopist'.[95] While atomic theory worked well in the case of simple molecules, however, Snow remarked glumly that in more complicated cases 'our ignorance of the real electronic states is likely to remain profound' – though in a note added to the article in proof he confessed that he may have been 'unnecessarily pessimistic' on this point, and he now saw it as 'likely that there will be a physical invasion of this side of chemistry within the next year or two'.[96]

Ultimately, for Snow, it was the explanation of chemical processes 'upon the basis of quantum physics' that represented 'the only satisfactory explanation of any atomic or molecular change'. Indeed, 'the quantum theory is so completely established in physics that the *translation* of other theories into its terms is desirable, even if nothing seems to be gained in the process; for the ideal to be reached is, of course, one body of facts which we now call "physical" and

92 C.P. Snow, 'Chemistry', in H. Wright (ed.), *University Studies: Cambridge 1933* (London, 1933), 97–121: 98–9.

93 Ibid., 103.

94 Ibid.

95 Ibid., 104.

96 Ibid., 106.

"chemical" and so on, all explained on the basis of the quantum laws'.[97] This assessment came at the moment of the abandonment of Snow's scientific career and his shift to a writer's vocation. It is all the more interesting, then, that he carried his admiration for the reductionist project across into one of his first literary ventures. In his 1934 novel *The Search*, the story of a promising young scientist who made a career-blighting mistake and took up a career as a writer instead, Snow's protagonist overhears 'one of the greatest mathematical physicists say, with complete simplicity, "Of course, the fundamental laws of physics and chemistry are laid down for ever. The details have got to be filled up: we don't know anything of the nucleus; but the fundamental laws are there. In a sense, physics and chemistry are finished sciences"'. The speaker 'was Newton's successor', and it was 'two hundred years since Newton talked of our being in the search for knowledge like children who pick up pebbles on the beach'. Now, Snow's protagonist muses, science was 'not picking up pebbles from the beach any more; instead we knew how many pebbles there were, how many we had picked up, how many we should be able to pick up. They had found the boundary to our knowledge; some things would remain unknown for ever'.[98] Snow's character.

> comforted [himself] by thinking of all the biological sciences; there were hundreds of years' work before us there; but even there, soon the framework would be laid down, and we should be just filling in the details. The nucleus and life: those were the harder problems: in everything else, in the whole of chemistry and physics, *we were in sight of the end*.[99]

The trajectory of reductionist unificationism in the interwar period – from a favoured rhetoric of physicists to material for science's publicists and storytellers – is a revealing and unexpected one. In serving as a professional and a public ideal, and against a changing background rhetoric of internationalism, imperialism and nationalism, usually in tension with each other, the idea of epistemological unity created and initially sustained a discipline that quickly showed itself in practice to be underdetermined and incapable of producing the sure outcomes the rhetoric demanded. Only a shared material culture could, in practice, produce the kind of unity required by the ideal – and even then it could not be guaranteed, as numerous disputes in nuclear physics in the 1930s illustrate. In the end, however, it was the scientists who abandoned the rhetoric of reductionism – and unity – and their armchair supporters who picked it up and continued to develop it in new ways in the 1930s

---

97 Ibid., 116, 119.

98 C.P. Snow, *The Search* (Harmondsworth, 1965 [London, 1934]), 162–3.

99 Ibid., 163; emphasis in original.

and later.[100] Indeed, it is no coincidence that in later years, as they fought their different cultural battles in Cold War Britain, the science writer Crowther and the novelist Snow – and, for that matter, the scientist Andrade – were among the most zealous and influential writers on Rutherford, the Cavendish and the reductionist programme.[101] Crowther wrote extensively about the 'genius' Rutherford and other Cavendish figures, and became the official historian of the Cavendish on its centenary in 1974, cementing in place the historiography of interwar nuclear physics he had played an important part in constructing.[102] Snow produced pen portraits of Rutherford and other Cavendish figures in his fictional and non-fictional writings.[103] For the physicists, unification by experiment turned out to be a fiction; for the fiction writers and the propagandists, it was more real than reality.

---

100 There are affinities here to analyses of the 'certainty trough' in which trust in knowledge claims is greatest '[b]etween those very close to the knowledge-producing technical heart of programs, and those alienated from them or committed to opposing programs'. See D. Mackenzie, *Inventing Accuracy: A Historical Sociology of Nuclear Missile Guidance* (Cambridge, MA, 1990), 371.

101 R. Desmarais, 'Science, Scientific Intellectuals and British Culture in the Early Atomic Age, 1945–1956: A Case Study of George Orwell, Jacob Bronowski, J.G. Crowther and P.M.S. Blackett' (PhD diss., Imperial College, London, 2010), 89–108; G. Ortolano, *The Two Cultures Controversy: Science, Literature and Cultural Politics in Postwar Britain* (Cambridge, 2009).

102 J.G. Crowther, *British Scientists of the Twentieth Century* (London, 1952), 43–92; idem, *The Cavendish Laboratory, 1874–1974* (London, 1974).

103 In a 1967 essay on Rutherford, for example, Snow observed that at the 1923 BAAS meeting, Rutherford 'announced, at the top of his enormous voice: "We are living in the heroic age of physics"', and 'went on saying the same thing, loudly and exuberantly, until he died fourteen years later'. 'The curious thing was', continued Snow, 'all he said was absolutely true'. See Snow, *Variety of Men* (n. 2), 1. For a similar perpetuation of the heroic historiography of interwar physics, see C.P. Snow, 'The Age of Rutherford', *Atlantic Monthly* 202 (November 1958), 76–81; idem, 'Rutherford and the Cavendish', in J. Raymond (ed.), *The Baldwin Age* (London, 1960), 235–48; idem, *The Physicists* (London, 1981). For recent critiques of Snow, see D. Edgerton, 'C.P. Snow as Anti-Historian of British Science: Revisiting the Technocratic Moment, 1959–1964', *History of Science* 43 (2005), 187–208; T.L. Lewis, *C.P. Snow's* Strangers and Brothers *as Mid-Twentieth Century History* (New York, 2009); Ortolano, *The Two Cultures Controversy* (n. 101).

Chapter 5
Scientists of the world unite
Socialist internationalism and the unity of science

Geert Somsen

A distinguishing feature of unity of science pursuits in the interwar period was their explicit internationalism. As Peter Galison shows in Chapter 2 of this volume, comparable movements of both earlier and later periods carried their politics at a national level, be it that of the German nation or the American democratic state. But the Unity of Science Movement of Neurath, Carnap and Morris in the 1930s sought to rise above nationality and unite the people involved with it on a higher plane – hence their insistence that their prime vehicle be named an *International* Encyclopedia of Unified Science. Comparable projects at the time shared this outlook (see Chapters 4, 6 and 9, this volume), and indeed H.G. Wells called his very similar undertaking a *World* Encyclopaedia. Unifying knowledge was supposed to bridge nationalities as well as scientific disciplines.

From one point of view this scientific internationalism was nothing new. The idea that truth transcends geography and that those in the pursuit of knowledge are united in a supranational community goes back at least to the early modern ideal of the Republic of Letters. Since then the notion has been reiterated time and again, and it even acquired new strength in the period around 1900 when the international scientific community was institutionalised in a string of novel forms: the international scientific conference, the international union per discipline and the International Association of Academies (IAA) as an overall umbrella. The IAA embodied the same dual unification as the Unity of Science Movement and its kin: that of all branches of science and of all nations of the civilised world.[1]

From a more pertinent perspective, however, the interwar internationalism was fundamentally different from what came before and particularly from the internationalism that blossomed around the turn of the twentieth century. It was different in its relation to nationalism and the carriers of national prestige. It was different in its political leanings and in how it presumed science to relate to politics. And it was different in its conception of science itself: of what science was and what it was for. If some kind of scientific universalism underlay both

---

1 Brigitte Schroeder-Gudehus, 'Les Congrès Scientifiques et la Politique de Coopération Internationale des Académies des Sciences', *Relations Internationales* 62 (1990), 135–48.

the old and the new view, the way this played out and the world order that this universalism represented differed massively between the two. What precisely these differences entailed and how and why they came about is the subject matter of this chapter.

Exploring such differences is a good way to reveal the political dimensions of scientific internationalism, which, at the face of it, seems a rather neutral and colourless affair. Actors have usually presented it as apolitical, in line with its scientific character, and internationalism in general appears to be a somewhat generic and unspecified political stance.[2] But by comparing the different forms it has taken and by tracing the changes in its underlying assumptions, we may break through these appearances and bring out its contrasting colours.[3] This will reveal the particular political orientations of scientific-internationalist initiatives and it will shed light on the international politics of interwar unity of science pursuits. Their internationalism, it will turn out, had a very specific agenda indeed.

The analysis here focuses on scientists from the Netherlands and their work in international organisations. This may seem like a severe limitation, but Dutch scientists were in fact among the most active internationalists in the interwar period, envisaging themselves as having a special mediating mission between the great powers.[4] Moreover, while they saw this task as unique, they took their conceptions of scientific internationalism from developments abroad and hence

2 Perry Anderson, 'Internationalism: A Breviary', *New Left Review* 14 (2001), 5–25.

3 See e.g. Jo-Anne Pemberton, 'The Changing Shape of Intellectual Cooperation: From the League of Nations to UNESCO', *Australian Journal of Politics and History* 58 (2012), 34–50; and Geert J. Somsen, 'A History of Universalism: Conceptions of the Internationality of Science, 1750–1950', *Minerva* 46 (2008), 361–79. Mark Mazower offers a genealogy of political internationalisms in *Governing the World: The History of an Idea* (London, 2012). Only few studies of scientific internationalism relate it to specific politics and conceptions of international order: Waqar Zaidi, 'Technology and the Reconstruction of International Relations: Liberal Internationalist Proposals for the Internationalisation of Aviation and the International Control of Atomic Energy in Britain, USA and France, 1920–1950' (PhD diss., Imperial College, London, 2008); '"A Blessing in Disguise": Reconstructing International Relations Through Atomic Energy, 1945–1948', *Past and Present* 210 (supplement 6) (2011), 309–31; '"Aviation Will Either Destroy or Save Our Civilization": Proposals for the International Control of Aviation, 1920–45', *Journal of Contemporary History* 46 (2011), 150–78; and Brigitte Schroeder-Gudehus, 'Probing the Master Narrative of Scientific Internationalism: Nationals and Neutrals in the 1920s', in Rebecka Lettevall, Geert Somsen and Sven Widmalm (eds), *Neutrality in Twentieth-Century Europe: Intersections of Science, Culture, and Politics after the First World War* (New York, 2012), 19–42.

4 W. Otterspeer and J. Schuller tot Peursum-Meijer, *Wetenschap en Wereldvrede. De Koninklijke Akademie van Wetenschappen en het Herstel van de Internationale Wetenschap Tijdens het Interbellum* (Amsterdam, 1997). Geert Somsen, '"Holland's Calling": Dutch Scientists' Self-Fashioning as International Mediators', in Lettevall et al., *Neutrality* (n. 3), 45–64.

their expressions reflect more widely shared views. The internationalism they began to voice in the 1930s was closely connected to other unity of science movements discussed in this volume, so the Dutch story will also shed light on them and vice versa. I will begin this chapter by characterising the dominant pre-war form of internationalism and by following its expression through the 1920s. I will then contrast it with a new type that arose in a number of initiatives in the 1930s and relate the latter internationalism to local Dutch as well as foreign contexts. In particular its relation to the British Social Relations of Science Movement will show it as integrated in a wider scientific-socialist programme. An epilogue traces its fate in the early Cold War.

The Olympics of science

One of the most striking features of the internationalism of Dutch scientists at the turn of the twentieth century is the extent to which they combined it with expressions of nationalism. Dutch science was very successful in the decades around 1900, and its accomplishments were increasingly framed in terms of national prestige. The Nobel Prize winning work of J.D. van der Waals, H.A. Lorentz and others was hailed as a 'physics of the fatherland', while the (similarly awarded) achievements of J.H. van 't Hoff and associates were celebrated as 'Dutch chemistry'.[5] The whole flourishing period itself was given a nationalist spin when a student christened it the 'Second Golden Age', suggesting a repetition of the seventeenth-century 'Golden Age', the heyday of the nation.[6] Nationalism was also expressed in a fervour for the history of Dutch science, in statues for great scientists of the past like Herman Boerhaave and in speeches at the *Nederlandsch Natuur- en Geneeskundig Congres* (NNGC, Dutch Congress of Science and Medicine), a Dutch version of the *Gesellschaft Deutscher Naturforscher und Ärzte*.[7] Its spokesmen not only trumpeted the national importance of scientific work, they portrayed it as a manifestation of the national character itself. According to NNGC president B.J. Stokvis, Dutch

---

[5] J.L. Oosterhoff, 'De Opkomst van een "Vaderlandsche Natuurkunde" aan de Leidse Universiteit in de Tweede Helft van de Negentiende Eeuw', in W. Otterspeer (ed.), *Een Universiteit Herleeft. Wetenschapsbeoefening aan de Leidse Universiteit vanaf de Tweede Helft van de Negentiende Eeuw* (Leiden, 1984), 114–17. Ch.M. van Deventer, 'Nederlandsche Scheikunde', *De Nieuwe Gids* 2 (1887), 144–7.

[6] K. van Berkel, *Citaten uit het Boek der Natuur. Opstellen over Nederlandse Wetenschapsgeschiedenis* (Amsterdam, 1998), 149–87: especially 149–50.

[7] Ibid., 221–39. R.P.W. Visser, 'Het "Nederlandsch Natuur- en Geneeskundig Congres" over de Relatie Natuurwetenschap en Samenleving, 1887–1900', in J.J. Kloek and W.W. Mijnhardt (eds), *Balans en Perspectief van de Nederlandse Cultuurgeschiedenis. De Productie, Distributie en Consumptie van Cultuur* (Amsterdam, 1991), 37–48.

people were by nature patient, dispassionate, precise and persevering, i.e. ideally equipped for scientific research. Dutch science expressed Dutch nationhood.

Little was original about this kind of nationalism – it very much reflected developments abroad. In France, science had counted as a manifestation of 'la grande nation' ever since Napoleon; its version of the NNGC carried the slogan 'Par la science, pour la patrie'.[8] German nationalism had always been built on notions of *Kultur*, of which *Wissenschaft* was a prime constituent. Scientists saw themselves as bearers of this culture and hence as representatives of the essence of Germankind. Their achievements, especially in pure science, were to be valued as cultural contributions, comparable to works in music, philosophy and art. As cultural goods they expressed the greatness of the nation, a greatness which, in turn, also rested on military strength and economic power.[9] This was not just a German, but a generally European understanding of scientific nationalism.

Rather than being opposed to it, such notions typically went hand in hand with scientific internationalism, including its new manifestations in the form of conferences, international unions and the IAA.[10] After all, national achievements in science could only be credibly assessed if measured against accomplishments elsewhere – that is: in international comparison and by international standards. Nationalism needed international frameworks and nationalists embraced them. Paul Forman has aptly compared this form of internationalism to what is displayed at the Olympic Games, where nations demonstrate their athletic strength by competing with other nations.[11] At the same time they celebrate this competition itself. The Games are as much about national triumph as about international fraternity, as much about battle as they are about peace. Similarly in science around 1900: nationalism did not clash with internationalism; the two facilitated each other.

---

8 Lorraine Daston, 'Nationalism and Scientific Neutrality under Napoleon', in Tore Frängsmyr (ed.), *Solomon's House Revisited: The Organization and Institutionalization of Science* (Canton, 1990), 95–115.

9 Russell McCormach, 'On Academic Scientists in Wilhelmian Germany', *Daedalus* 103 (1974), 157–71. On the role of science in the national unification of Germany, see Chapter 2, this volume. On scientific nationalism generally, see Ralph Jessen and Jakob Vogel (eds), *Wissenschaft und Nation in der Europäischen Geschichte* (Frankfurt, 2002).

10 Brigitte Schroeder-Gudehus, *Les Scientifiques et la Paix. La Communauté Scientifique Internationale au Cours des Années 20* (Montréal, 1978), 41–62; Schroeder-Gudehus, 'Congrès scientifiques' (n. 1), 137; Gabriele Metzler, *Internationale Wissenschaft und Nationale Kultur. Deutsche Physiker in der internationalen Community 1900–1960* (Göttingen, 2000) 33–83.

11 Paul Forman, 'Scientific Internationalism and the Weimar Physicists: The Ideology and Its Manipulation after World War I', *Isis* 64 (1973), 151–80: 154.

The situation in the Netherlands in the early twentieth century perfectly fit this Olympic model. Dutch scientists combined ardent nationalism with great enthusiasm for international organisations and joined the various international unions and the IAA as soon as they could. But there was also a more particular trait to Dutch internationalism. Some scientists claimed that the Netherlands should not merely adhere to such organisations but offer to provide them their permanent seat. The physician P.H. Eijkman, for example, proposed to bring the headquarters of all international scientific unions to The Hague, where they would be accommodated in a gigantic 'World Capital' surrounding the new Peace Palace (which, he thought, had a similar function).[12] Not everybody took Eijkman's grandiose plans seriously, but they did resonate with a growing ambition to make the Netherlands the home of international institutions. In 1913, the Amsterdam Academy attempted to get the IAA based in a Dutch city. At an IAA meeting in St-Petersburg, it proposed to establish a permanent secretariat, with an eye on bringing the offices of that secretariat to the Netherlands. The German and Austrian academies blocked this proposal but this did not stop Dutch scientists' desire to have their country play a central, organising role in the world of science.[13]

In this ambition nationalism and internationalism were not just (Olympically) combined, they essentially merged. For not only did the Dutch hail national triumphs at the same time that they praised international cooperation. They also took national pride in facilitating this international cooperation, a task that would become known as 'Holland's Calling'. As professor of medicine H. Burger said, in support of Eijkman's plans:

> [T]he international movement [of organising conferences etc.] . . . involves a national interest that shouldn't be underestimated. This internationalism is by no means at odds with the defence of our national autonomy. On the contrary, it is our patriotism that should spur us to ensure that our country is mentioned as honourably as possible among the nations.[14]

Just as it is not only prestigious to win at the Olympic Games but also to organise them, in the world of science the *organisation* of international exchange

12 Geert J. Somsen, 'Global Government through Science: Pieter Eijkman's Plans for a World Capital', in Boyd Rayward (ed.), *Information Beyond Borders: International Cultural and Intellectual Exchange in the Belle Époque* (Aldershot, 2014), 201–19.

13 Otterspeer and Schuller, *Wetenschap en Wereldvrede* (n. 4), 49–50. The official name in this period was Royal Academy of Sciences, but I will use the informal 'Amsterdam Academy' for the sake of geographical specificity.

14 H. Burger, 'Wetenschappelijk Internationalisme', *Nederlandsch Tijdschrift voor Geneeskunde* 52, 2nd half (1908), 216. This and all subsequent translations to English are mine.

was a matter of national prestige as well. For this reason, recurring international conferences often circulated, so as to ensure that every country got its share in the honour of hosting. But harbouring more permanent institutions or mediating between the greater powers are tasks that cannot pass from country to country. They have often been taken up by small nations like Belgium, Switzerland, Sweden and the Netherlands.[15] Here then is the sense in which Dutch internationalism was different from that of most nationalities: it was still Olympically conceived but had the extra element of an organising ambition. In 1913, the Amsterdam Academy had failed in its attempt to seize this role in the IAA. But the Great War would produce new opportunities which Dutch scientists would pick up with the zeal of a national vocation.

Mediations between nations

The possibility for Dutch mediation arose from the rift in scientific relationships between the great powers after the war. In 1919 formerly Allied scientists replaced the IAA by the International Research Council (IRC), an organisation with the stated aim of excluding the Central Powers from membership.[16] The exclusion was a consequence of the infamous *Krieg der Gelehrten*, the war within the war in which scientists of both sides had morally defended their own countries' military pursuits and come to question (and finally denigrate) the trustworthiness of their counterparts 'across the Rhine'. German academics' denials of their army's atrocities in Belgium had especially created much bad blood among Allied scholars and, like the League of Nations, the IRC chose to temporarily prohibit German membership.[17] It did, however, invite academies from neutral countries to join the Allied club.

The Dutch were among these invitees but they hesitated. Their internationalist views did not readily square with the new organisation and in a series of Academy meetings they hotly debated the pros and cons. On the one hand,

---

15 On small nation internationalism, see Madeleine Herren, *Hintertüren zur Macht. Internationalismus und modernisierungsorientierte Aussenpolitik in Belgien, der Schweiz und den USA, 1865–1914* (München, 2000) and Rebecka Lettevall, Geert Somsen and Sven Widmalm, 'Introduction', in Lettevall et al., *Neutrality* (n. 3), 1–15: 4–6.

16 Daniel J. Kevles, '"Into Hostile Political Camps": The Reorganization of International Science in World War I', *Isis* 62 (1971), 47–60. The IRC was only a continuation of the natural science part of the IAA. The humanities were reorganised into the Union Académique de Recherches et de Publications with the same exclusions.

17 Jürgen von Ungern-Sternberg and Wolfgang von Ungern-Sternberg, *Der Aufruf 'An die Kulturwelt!' Das Manifest der 93 und die Anfänge der Kriegspropaganda im Ersten Weltkrieg* (Stuttgart, 1996).

most academicians questioned the 'universally international' nature of the IRC.[18] Having not suffered as much during the war, they did not share the Allies' anti-German sentiments. To them excluding the most important scientific nation on political grounds looked like a violation, rather than a continuation, of scientific internationalism. On the other hand, *if* the Academy joined the IRC, it could try to change this situation from within, assuming the role of a neutral bridge builder and mediator. In that sense, the post-war boycott gave Dutch scientists a golden opportunity to live up to 'Holland's Calling' as organisers of true international cooperation. This argument finally prevailed, and the Academy decided to enter the IRC.[19]

From the beginning of their membership Dutch scientists indeed made great efforts to accomplish international reconciliation. The Academy even established a special committee 'for achieving everybody's ideal: general cooperation of *all* nations in the realm of science'.[20] Only at one point did it officially propose that the IRC change its statutes and allow Central states' membership (in 1925, together with the Danish Academy, and without success). But throughout the 1920s Dutch academicians were working behind the scenes, monitoring the moods of their foreign colleagues and trying to find the right moment to take a next step. H.A. Lorentz, the *éminence grise* of Dutch physics, especially developed into a true science diplomat, respectfully approaching his Allied and Central counterparts and always keen to find a formulation that all sides could accept. In 1926, the IRC did finally open its doors to the formerly excluded countries, but now it was the Germans who refused to join what they saw as a political and arrogant institution. And so the Dutch mission continued.[21]

Also outside the Academy, through personal contacts and in the international unions per discipline, scientists from the Netherlands worked at reunification. In 1921, the chemistry professor Ernst Cohen organised an informal meeting at his home in Utrecht with colleagues from France, Germany, Britain, Austria, Russia and Denmark (see Figure 5.1). This was followed by a bigger conference the next year, more formal efforts in the International Union of Pure and Applied Chemistry (whose president Cohen became in 1925) and finally by German entry

[18] 'De Internationale Wetenschappelijke Samenwerking Hersteld', undated typescript, file 73b, Archief Koninklijke Nederlandse Akademie van Wetenschappen, 64 (hereafter 'KNAW Archive'), Noord-Hollands Archief, Haarlem.

[19] The decision was far from unanimous, however. Almost one quarter of the academicians votes against entry and the prominent members G. Heymans and J.C. Kapteyn left the Academy for it. See Otterspeer and Schuller, *Wetenschap en Wereldvrede* (n. 4), 95–118 and Somsen, 'Holland's Calling' (n. 3), 52–5.

[20] Minutes, 24 February 1923, Wetenschappelijke Internationale Samenwerkingscommissie, 538 (hereafter 'WIS'), KNAW archive (emphasis in original).

[21] Otterspeer and Schuller, *Wetenschap en Wereldvrede* (n. 4), 119–204.

Figure 5.1 Meeting of chemists from 'across the Rhine' at Ernst Cohen's Utrecht home in 1921 (picture taken in his garden). These participants were all close acquaintances of Cohen's which helped in overcoming resistance – only one invitee, from Belgium, declined. Cohen sought to expand participation at a larger meeting the following year, but there both French and Belgian chemists chose to stay away. From left to right: Pieter van Romburgh (Netherlands), Rudolph Wegscheider (Austria), Hugo Kruyt (Netherlands), Paul Walden (Germany), Victor Henri (France), Rudolf Schenck (Germany), Ernst Cohen (Netherlands), Einar Biilmann (Denmark) and Frederick Donnan (Britain).

Source: Ernst Cohen, *Hugo Rudolph Kruyt 1908–1933. Ter Herdenking van zijn 25-Jarig Doctoraat* (Amsterdam, 1933).

into that union in 1930.[22] Similar actions were taken by Willem de Sitter, who got Germans to participate in an International Astronomical Union convention in Leiden, and F.A.F.C. Went, who was working on internationalisation in biology. Lorentz used his position of chairman at the Solvay conferences in physics and at the League of Nations's *Commission Internationale de Coopération Intellectuelle* to re-establish all-inclusive participation.[23]

22 Jorrit Smit, 'Nuclei in a Supersaturated Solution. Utrecht Chemists and the Crystallization of International Relations after the First World War', *Studium* 7 (2014), 190–208. Ernst Cohen, 'Hugo Rudolph Kruyt 1908–1933', in *Hugo Rudolph Kruyt 1908–1933. Ter Herdenking van zijn 25-Jarig Doctoraat* (Amsterdam, 1933), 28–35.

23 Otterspeer and Schuller, *Wetenschap en Wereldvrede* (n. 4), 125–33, 136–8.

The Dutch met with great appreciation for their mediating efforts, but also with resistance. For some Allied scientists, especially from France and Belgium, the mediations simply came too early. The memories of German atrocities, and of their denial by German professors, were still fresh and made it hard to resume normal cooperation. These scientists felt that the Dutch were insensitive in their attempts to return to business as usual and act as if nothing happened. When after 1926 the counter-boycott reversed the situation it was German scientists who believed that the Dutch were inconsiderate and failed to see how humiliatingly the IRC was treating them. Now *they* resented mediation. Reconciliation might be nice, but there were reasons for the rift and these could not be ignored.[24]

Yet even if the reactions were mixed, it was the positive responses and gratitude that dominated the way the Dutch experienced the reception. Cohen observed mainly goodwill among the participants of his conferences. W.P. Jorissen, editor of the *Chemisch Weekblad* (*Chemical Weekly*), noted how most foreign chemistry journals published favourable reports of these meetings.[25] Cohen himself quoted a poem that the chemist Paul Walden (of Latvian-Russian-German extraction) had written on the occasion, comparing the attempts to reconcile the belligerents to the Dutch tradition of taming the sea.[26] In their reports, the Dutch often mixed praise for mediation with respect for the status of their science. According to De Sitter, there was 'no doubt that our position between the nations . . . is powerfully supported by the esteem for our scientific achievements.'[27] Both he and Went heard at conferences that 'in science we count among the great powers'.[28] The *Chemisch Weekblad* made a similar point citing a remark by the American chemist W.A. Noyes:

> We [Americans] were particularly impressed by a statement [Cohen] made to us about the Nobel prizes. You have here some five million people . . . but he told us that you have received in Holland five Nobel prizes for scientific work. We in America have somewhat more than a hundred million people and we have received two prizes for such work. I am sure that no other country can show such a record as Holland.[29]

No doubt Noyes was sincerely impressed but note how his praise was both elicited and reported by Dutchmen themselves. International admiration was eagerly absorbed and turned into a source of national pride.

---

[24] See for example Kenneth Bertrams, 'Caught-up by Politics? The Solvay Councils on Physics and the Trials of Neutrality', in Lettevall et al., *Neutrality* (n. 3), 138–58: 149–51.

[25] Smit, 'Nuclei' (n. 22), 202.

[26] Cohen, 'Kruyt' (n. 22), 32.

[27] 'Rede van Prof. Dr. W. de Sitter', *Fysica* 6 (1926) 1–6: 3.

[28] Ibid., 2, and Otterspeer and Schuller, *Wetenschap en Wereldvrede* (n. 4), 138.

[29] W.P. Jorissen, 'Réunion Internationale de Chimie à Utrecht, 21–24 Juin 1922', *Chemisch Weekblad* 19 (1922), 418–425: 422.

But these reports were not just ego-massage. As De Sitter's quotation already suggested, foreign respect also played a role in how the Dutch justified that it was their task to mediate between the greater nations. This justification hinged on a distribution of national greatnesses. Militarily and economically the Netherlands was relatively weak. But scientifically it could claim great power status – at least that is what the praise and remarks such as Noyes's seemed to confirm. It was precisely this combination of qualities – economic and military weakness on the one hand and scientific grandeur on the other – that served to legitimate the mediating role. Since the Netherlands formed no geopolitical threat to any nation, its scientists could not be suspected of ulterior motives. But because it was scientifically strong, they could speak and act with authority. The international respect that the Dutch experienced hence served as a foundation for their mediating role.[30]

Despite the turbulence created by the Great War then, the internationalism enacted by Dutch scientists in the 1920s was still of the pre-war, Olympic kind. It was an internationalism that was deeply connected to nationalist sensibilities. It was an internationalism that took the nation as the unit of interaction (literally 'inter-nationalism'). And it was an internationalism that measured the status of nations in terms of comparative scientific accomplishment. For this reason the participation of countries like Morocco or Siam was sometimes frowned upon – Went, for instance, complained about the influence of 'uncivilised and half-civilised countries' in the IRC.[31] Conversely, the argument of scientific achievement provided the Dutch with a justification of their role as mediators. One difference with the internationalism before the war was that the peace-promoting effects of science were less heavily emphasised: in the 1920s, the scientific community was itself in need of pacification. But the general conception was the same. The internationalism of the 1920s was still one that spurred rather than countered nationalism, and that regarded science as a cultural good, which – just like economic and military strength – formed an expression of the greatness of a nation.

Socialist internationalism

But things were to change. In the 1930s a different kind of internationalism came to inform the pronouncements and actions of the Dutch, reflecting, as we shall see, developments in other countries. This was not because international relations in science themselves had become drastically different. It is true that the transformation of the IRC into the International Council of Scientific Unions (ICSU), in 1931, shifted power from the adhering national organisations to the

30 Lettevall et al., 'Introduction' (n. 15), 4–6.

31 Quoted in Otterspeer and Schuller, *Wetenschap en Wereldvrede* (n. 4), 146.

disciplinary unions. But this had no effect on the membership situation.[32] The former Allied countries still all adhered to the ICSU and the Central Powers were still welcome to join but did not. When after 1933, the national socialists started to control German academic policy, international reconciliation was even further away. If anything, the need for mediation seemed to become only bigger and the Dutch scientists continued on their mission.

In April 1934, the Amsterdam Academy of Sciences submitted a proposal to the ICSU to be adopted by its general assembly, which was to meet in Brussels in July of the same year. The proposal asked the ICSU to issue a statement, preformulated by the Dutch academicians, calling upon scientists to help remove the growing barriers between the various nations. In some respects, the proposal was a continuation of the earlier mediating efforts. The statement text urged all scientific organisations to maintain international cooperation and to include as many countries as possible – especially in these times of increasing political tensions. The text even revived some of the old pacifist rhetoric that had taken a back seat during the 1920s. It stressed 'the possibility and necessity of peace between the world's peoples' and stated that 'the "brotherhood of scientists" can be an important factor' toward reaching that end.[33] As ICSU President N.E. Nörlund said when he presented the proposal to the general assembly: '[I]n times of multiplying barriers between the peoples . . . it is important to set the example of disinterested scientific work'.[34]

But the statement also showed some striking new features. For one, it was full of economic considerations. The text described the present crisis not just in terms of growing tensions between states but also strongly associated it with problems of industry, employment, housing, nutrition and so forth. Scientists, it said, would no doubt play a major role in tackling these issues until ultimately 'a more harmonious economic structure' would be found – ideally by 'an international organization devoted to the solution of these problems'.[35] At this moment, however, nations tended to deal with the difficulties each on their own and therefore 'the scientists of every country will be drawn more and more into . . . spheres of national organization.' It was this tendency that the Academy proposal warned against. Scientists might be called upon to help out in their particular countries, but while doing so they should never lose sight of their higher

---

32 Frank Greenaway, *Science International: A History of the International Council of Scientific Unions* (Cambridge, 1996), ch. 3.

33 'Proposal by the Royal Academy of Sciences, Amsterdam', in H. Lyons, (ed.), *The Second General Assembly of the International Council of Scientific Unions held at Brussels, July 9th to 13th, 1934. Reports of Proceedings* (London, 1935), 16–17: 17.

34 N.E. Nörlund, 'Discours du Président', in ibid., 11–15: 11.

35 'Proposal' (n. 33), 16.

obligations. They should continue to operate as an international brotherhood, and they should combat 'the dangers involved in a too exclusive nationalism'.[36]

The entire tenor of the statement differed from the earlier internationalist discourse. Any talk of 'harmonious economic structures' or practical social problems had been quite alien to that. The stress used to be on pure science and its cultural significance, but now science was discussed primarily in terms of its problem-solving capacity – the 'relations between pure and applied science' were even explicitly emphasised.[37] The new view also suggested that it was the *organisation* of science that made it become a national endeavour, whereas before, the pursuit of science itself had been a part of national culture, almost by definition. Moreover, these national dimensions were now depicted as an evil that needed to be countered by international cooperation. Internationalism no longer embraced nationalism, it was pitted against it. Overcoming 'national tendencies' was the main task of the 'brotherhood of scientists'.

The new direction was taken even further in a proposal that the Amsterdam Academy submitted to the next ICSU general assembly, to be held in London, in 1937. This time transcending national boundaries was no longer considered to be the only task of international science, it was part of a much larger set of 'social responsibilities'. The background to the proposal, an accompanying memorandum stated, was a growing concern about international developments such as the military build-up in Germany, the rearmament of other countries and the devastating economic depression. What the political and economic crisis showed, the authors claimed, was that the current governments and industrial powers were unable to analyse and solve problems in a way that promoted the well-being of humanity as a whole. Instead, they increasingly withdrew in their national spheres and headed for international conflict. The press and radio were not working for the 'unity of mankind' either – they issued nationalist propaganda, appealed to base, tribal emotions and spread 'mental and moral confusion' among the masses. All trends seemed toward disunity. Fortunately, however, the proposal authors noted, there were 'thinking people, and in particular scientific workers', who recognised this as the wrong way to go. Scientists were internationalists at heart and they knew that cooperation was always the more rational alternative.[38]

The text claimed that under these circumstances scientists had a number of special duties. Their first task was to fight the current 'tendency towards war'.

36 Ibid., 17.

37 Ibid., 16.

38 'Memorandum Concerning a Proposal for the Appointment of a Committee for Investigating the Social Responsibilities of Science and of Scientific Workers, to be Submitted to the Meeting of the International Council of Scientific Unions at London in April 1937', 2–4, typescript, Scientific Unions, file 455 (hereafter 'SU'), KNAW archive.

Leaders of government and entire populations seemed to proceed wilfully toward armed conflict, and this 'most dangerous and contagious mental disease' should be eliminated like a bacterial epidemic. As guardians of truth and rationality, scientists also had a role to play. They should 'analyze what is misleading and false in . . . propaganda' and mobilise a moral force against the 'temptation towards . . . barbaric means of action'.[39] Half-truths should be replaced by correct information and appeals to tribal instincts by 'an education preparing for sound thinking'.[40] Scientists should impede material war preparations as well. They should refuse to assist in the production of chemical or biological weapons and instead work on the protection of civilians and the development of international defence plans. Finally, the memorandum discussed the attitudes of scientists toward the military. If treated like civil organisations, such as the police or fire brigades, the armed forces could do good work. But if they were allowed too much power, the military would act in ways that were incompatible with the values of science. They would demand secrecy from researchers, for instance, and they would try to control public opinion – both in conflict with the spirit of openness and freedom of thought that science stood for. Generally speaking, the military favoured brute force and patriotism over argument and reason, creating 'a fighting spirit which . . . irresistibly forces men to have recourse to acts of barbarism.'[41] Militarism was the antithesis of the ethos of science.

The tone was firmer, the message sharper and the range of issues much broader than before. This proposal was no longer just about the 'brotherhood of scientists' advancing international cooperation but about the significance of a scientific attitude for a whole series of phenomena – from militarism and propaganda to public awareness and biological warfare. Still all these evils were effectively rooted in one main development, the same development that the 1934 proposal had responded to. This was the tendency of politicians and populations to withdraw into spheres of national organisation and to fall back on nationalist ideologies. This tendency had been criticised in 1934, but was now condemned in much stronger terms. Nationalism was not just opposed to internationalism, it was associated with irrationality, with 'unsound creeds and prejudices' and ultimately with barbarism.[42] Its current growth was a retrograde development, contrary to the direction of scientific progress and incompatible with anything that science stood for. How far these notions differed from the views presented in the 1920s is perhaps best illustrated by comparing the perceptions of the relation of science to the military. In the older view, military

39 Ibid., 4.

40 Ibid., 2.

41 Ibid., 4.

42 Ibid., 4–5.

strength was regarded as something on a par with scientific accomplishment – both being expressions of the greatness of a nation. But in the new conception, the two were completely contrary to each other, representing values and moving in directions that were diametrically opposed. As the memorandum stated,

> The old historical tendency which considered the military and the naval organization as the most important factor of a nation and as the prime expression of its force, should nowhere find support among scientific workers, as it is contrary to our present idea of cultural advance.[43]

Compared to Olympic internationalism the reversal of values was complete.

Dutch debates

The question is why this change took place. What made one kind of scientific internationalism give way to another that was conceived in an entirely different way? One of the most obvious and general reasons was, of course, the transformation of the international political scene that had occurred since the 1920s. The aftermath of the First World War had made place for a new set of international tensions, spurred by the great Depression and the growing menace of Nazi Germany in Europe. Under these circumstances it no longer made sense for Dutch scientists to continue their silent diplomacy or even to try to reinvolve their German colleagues in the international organisations. But while this may explain the end of their previous type of mediating efforts, it does not account for the particular new form that their internationalism took. For that we have to look at developments within the Dutch scientific community.

What had also changed between the 1920s and 1930s was the identity of the internationalisers in the Amsterdam Academy. Lorentz, the undisputed leader, had died in 1928. His successor as chairman, Went, passed away in 1935 and was already in too fragile health to attend the ICSU meeting the year before. Cohen remained active in the Academy, but he was not behind the proposals of 1934 and 1937 – in fact, he had raised objections against them, as had Went (in 1934) and a number of other older academicians.[44] The people initiating and backing the two proposals tended to be younger scientists who had worked in the second rank during the 1920s and were now moving to leading positions in the Dutch scientific establishment. They were just as convinced internationalists as the generation before them but on a different basis.

---

43 Ibid., 4.

44 Minutes meeting 28 April 1934, Minutes of the Afdeeling Natuurkunde, file 17 (hereafter 'AN'), KNAW archive; minutes, 13 January 1937, WIS; minutes, 13 January 1937, AN.

Most important in this group were two scientists who had been the driving forces behind the 1934 and 1937 proposals: the Utrecht chemistry professor H.R. Kruyt (1882–1959) and the Delft professor of aero- and hydrodynamics J.M. Burgers (1895–1981).[45]

Both Kruyt and Burgers had been active internationalisers from early on in their careers. Kruyt had been among the first to represent the Netherlands in the International Union of Pure and Applied Chemistry (IUPAC) when it was founded in 1919. He had also assisted Cohen with the international chemists' meetings in the early 1920s and chaired such a conference in 1928. He became vice president of IUPAC from 1928 to 1934 and of the ICSU from 1937 to 1945. After the Second World War, he would also be president of both and sit on the board of UNESCO.[46] Burgers had been involved in international affairs from early on too. Soon after he switched to aero- and hydrodynamics (he had been trained as a theoretical physicist in Leiden), he organised an international conference in that field in 1924. This meeting was to be the first of a regular series that eventually led to the foundation of the International Union for Theoretical and Applied Mechanics (IUTAM) in 1946. Burgers authored its statutes and became its first secretary. In 1937, he was also appointed secretary to the ICSU's Committee on Science and its Social Relations (more about this later).[47]

In much of this activity Kruyt and Burgers followed in the footsteps of a previous generation, but they differed in the extent to which they combined their internationalism with explicit political positions. Both were overt socialists. Kruyt had developed his left-wing orientation during his college years, when what was known as the 'socialist flood wave' swept the Amsterdam student population and he became deeply involved in political debates. He combined his socialism with Christian belief and continued to proclaim

---

[45] Kruyt and Burgers appear as the authors of the 1937 proposal. The 1934 proposal was officially authored by the Academy's International Scientific Cooperation Committee as a whole, but Burgers probably wrote at least the final version. As the committee's secretary, he submitted the typescript to the board of the Academy's science department. He and Kruyt and the Delft botanist G. van Iterson defended it at the ICSU meeting. Other support in the Academy came from Utrecht physicist L.S. Ornstein and the physicians J. van der Hoeve and B. Brouwer, chairmen of the science department, in 1934 and 1937, respectively. The strongest opponents (besides Cohen and Went) were the mathematicians L.E.J. Brouwer and R. Weitzenböck.

[46] Geert Jan Somsen, *'Wetenschappelijk Onderzoek en Algemeen Belang'. De Chemie van H.R. Kruyt (1882–1959)* (Delft, 1998); J.Th.G. Overbeek, 'Bibliografie H.R. Kruyt', *Chemisch Weekblad* 54 (1958) 322–24: 324.

[47] Fons Alkemade, 'Biography', in F.T.M. Nieuwstadt and J.A. Steketee (eds), *Selected Papers of J.M. Burgers* (Dordrecht etc., 1995), xi–lxxxvi: xiii–xx and xlviii–liii.

his Christian-socialism through public speeches and other fora for the rest of his life.[48] Burgers had also developed his political interests during his student years. He took part in lively discussions with his friends Dirk Struik, the future mathematician *cum* communist *cum* historian of science, and Jan Romein, who became a well-known left-wing historian. After the Bolshevik Revolution in Russia, Burgers became convinced that socialism was the only rational way forward, and in 1918 he joined the *Communistische Partij Holland*. He left the party again 15 years later when he started to disagree with the dictates from Moscow, but his socialist leanings and his search for a new social order remained.[49]

In itself, a socialist orientation was not new in Dutch academia by the 1930s. But it was unusual to express such political convictions in academic activities, especially within the natural sciences. The 1934 and 1937 proposals, however, both contained phrasings that certainly could make that impression. All the talk of a dawning 'more harmonious economic structure', of 'necessary adjustments to new social structures' and of failing 'governing bodies and . . . industrial and commercial concerns' seriously smacked of socialism.[50] This caused controversy among Dutch scientists when the proposals were first discussed in the Academy meetings. In 1934, Cohen, Went and others worried that the expressions might damage the ICSU's political neutrality which was so easily questioned anyway.[51] And in 1937, L.E.J. Brouwer, backed by Cohen and others, claimed that the proposal was in fact a political statement which therefore required approval from the Dutch government (which, as he knew, was conservative). Even the astronomer Anton Pannekoek, who was a prominent communist himself, objected to the last proposal. Although he generally sympathised with its contents, he did not feel that the international *scientific* community should play a significant role 'in the current battle of ideologies'.[52] Like the

---

48 Somsen, *Wetenschappelijk Onderzoek* (n. 46), 172–7. Kruyt's brother John William was briefly MP for the radical Christian-Socialists and ended his life as a Russian spy in the Netherlands during Nazi occupation. See Herman Noordegraaf, *Het Christen-Socialisme van John William Kruyt en Truus Kruyt-Hogerzeil* (Voorburg, 1990).

49 Autobiographical sketch, Johannes Martinus Burgers papers, Internationaal Instituut voor Sociale Geschiedenis, Amsterdam. G. Alberts, 'On Connecting Socialism and Mathematics: Dirk Struik, Jan Burgers, and Jan Tinbergen', *Historia Mathematica* 21 (1994), 280–305: 281–3 and 293–8.

50 'Proposal' (n. 33), 16; 'Memorandum' (n. 38), 2 and 5.

51 Minutes, 28 April 1934, AN.

52 Brouwer and Pannekoek voiced their objections in the meeting of the International Scientific Cooperation Committee. Cohen and others did so in the meeting of the Academy's science department, that took place the same afternoon. See minutes, 13 January 1937, WIS; and minutes, 13 January 1937, AN.

other opponents, Pannekoek preferred to keep his political activities apart from his scientific work – even to the extent that, later in life, he would write two separate autobiographies.[53]

The supporters of the proposals had to make great efforts to get them passed. In 1934, it was still enough for Kruyt to promise his Dutch colleagues that he would inform the ICSU meeting of the objections that some of them had raised. But in 1937, Kruyt and Burgers needed to affirm repeatedly that the proposal had no political intentions. They could maintain this because they had split their text in a 'Proposal' document and a 'Memorandum'. All the explicit passages cited above were part of the Memorandum which was itself not up for vote. The Proposal text was, but it only called for the installation of a committee which would collect opinions on the scientific standpoint in certain international developments. And in fact even the Memorandum only meant to present examples of such opinions. Even so, the Dutch Academy members sensed the general direction of the initiative, and a majority of them were only prepared to send the proposal to the ICSU on condition that Burgers would add an 'Explanatory Note' that the intentions were entirely apolitical and that Kruyt would confirm this verbally when he delivered the proposal.[54]

All the same, the discussions repeated themselves at the ICSU general assembly. In 1934 the Academy proposal had still been quite easily accepted.[55] But in 1937 the Dutch initiative became the most heavily debated issue at the meeting, with 15 different representatives engaging in what the official report called an 'active discussion'. The general reservations were, again, that the proposal might jeopardise the ICSU's neutrality, that it went beyond the ICSU's proper domain and that 'political movements' might start to influence science. Kruyt then withdrew the proposal, while an ad hoc group of ICSU delegates quickly drafted a new version. This alternative proposal was eventually passed, but only because its scope had been seriously reduced. In the new plan, the proposed committee was merely supposed to prepare regular surveys of recent scientific results, their effect on the world picture and possible practical

53 David Baneke, '"Hij Kan Toch Moeilijk de Sterren in de War Schoppen". De Afwijzing van Pannekoek als Adjunct-Directeur van de Leidse Sterrenwacht in 1919', *Gewina* 27 (2004), 1–13.

54 Minutes, 13 January 1937, WIS; minutes, 13 January and 27 February 1938, AN.

55 G. van Iterson reported that the ICSU Executive Committee had raised no objections, even after Kruyt had informed them about the reservations of some Dutch academicians. The proposal was then passed on to the General Assembly, which adopted it unanimously. G. van Iterson to B. Brouwer (the Academy secretary), 12 July 1934, SU. See also 'Proposal' (n. 33), 17.

applications. 'The work of the Committee', it was added to the task description, 'is limited strictly to scientific activity'.[56]

Still, the Dutch supporters of the initiative were quite happy with this result. One of the delegates reported that the undertaking had been successful and that the ICSU had adopted the proposal in a 'slightly altered formulation'.[57] And Burgers agreed to serve as secretary to the new committee, despite its enormously reduced scope. It would soon become apparent that this reduction did not have much effect anyway. As the 'Committee on Science and its Social Relations' (CSSR), the group started to gather viewpoints on the social role of science much along the lines of the original proposal. In the summer of 1938, they sent out a questionnaire to correspondents all over the world, asking them about exactly the sorts of topics that had been outlined in the Memorandum: Is the public protected from propaganda and its 'spreading of contagious mental aberrations'? Are governmental, industrial and commercial institutions in your country able to deal with problems efficiently? Does your country demand 'loyalty and service either to the nation, or to the whole of mankind'?[58] The CSSR's questions made it look as if the debates at the ICSU general assembly had never taken place. And the report on the social relations of science that was to result from the questionnaire (scheduled for publication in 1940) promised to be something very close to what Kruyt and Burgers had had in mind.

How was this possible? How could the CSSR get away with ignoring the dominant opinion of the assembly that had established it? Partly, because of sneaky tactics. The members of the CSSR were not appointed by the general assembly, but by the ICSU executive board, three months after the conference in London. And the new executive board was dominated by people who were warmly sympathetic to the original proposal: Kruyt himself was its vice president, and Cambridge astrophysicist F.J.M. Stratton was a member who would become president of the CSSR.[59] But it was not just this strategy that enabled the continuity. There was also a conviction among the original

---

56 F.J.M. Stratton, *The Third General Assembly of the International Council of Scientific Unions, Held at Brussels, April 26th to May 3rd: Reports of Proceedings* (Cambridge, 1938), 17–20. G. van Iterson, 'Rapport van de Delegatie naar de Vergadering van de International Council of Scientific Unions te London van 26 April–3 Mei 1937', AN.

57 Van Iterson, 'Rapport' (n. 56).

58 *Report on the Work Done by the 'Committee on Science and its Social Relations' Instituted by the International Council of Scientific Unions* (typescript June 1938), SU. This report included the questionnaire.

59 Burgers and Stratton had also been part of the ad hoc committee that drafted the second proposal during the 1937 General Assembly.

supporters that the second version of the proposal *was* in essence identical to the first and that the original proposal had not contained the extraneous political elements that the general assembly had seen in it. Kruyt and Burgers had a different conception of the relation of science to politics – different from most delegates at the ICSU meeting and different from older Dutch academicians like Cohen and Pannekoek. This conception was central to the new brand of internationalism – and tracking its nature and origins will lead us beyond the Dutch situation and consider the international context of the initiatives.

British backgrounds

Besides resistance, the Dutch initiatives also received a lot of foreign support. The journal *Nature*, for example, hailed the 1934 proposal and applauded its attempt to present the model of international scientific cooperation to 'those who are politically in control of world affairs'.[60] In 1937 again, *Nature* carried a strongly supportive editorial, explicitly backing the action against 'retrograde political creeds and nationalist systems'.[61] Reprints of this article were handed out at the ICSU general assembly on the day of the proposal's discussion.[62] At the same time, the assembly attendants received a copy of the Memorandum, which Kruyt and Burgers had precirculated to colleagues at home and abroad, and which now carried 79 signatures of approval, most of them foreign. The list contained important names like Rutherford and Soddy from Britain; Langevin from France; Langmuir, Millikan and Pauling from the U.S.; and Establier of the League of Nations.[63] The most conspicuous group, however, consisted of British scientists representing what has become known as the Social Relations of Science Movement: the radical socialists J.D. Bernal, Joseph Needham, P.M.S. Blackett and Hyman Levy as well as the more moderate 'reformers' Julian Huxley, John Boyd Orr and Richard Gregory, the editor of *Nature*.[64]

---

60 'International Co-operation in Science', *Nature* 133 (1934), 961–2: 961. The fate of the proposal was also favourably reported in *Nature* 134 (1934), 89–90; *Science* 80 (1934), 132; and in the *Bulletin de l'Institut International de Coopération Intellectuelle de la Société des Nations* 43/44 (1934), 411–12.

61 'Social Responsibilities of Science', *Nature* 139 (1937), 689–91: 691.

62 Van Iterson, 'Rapport' (n. 56).

63 H.A. Kramers and others to Sir Henry Lyons, 31 March 1937, WIS.

64 On the distinction between 'radicals' and 'reformers' see Gary Werskey, 'British Science and "Outsider" Politics, 1931–1945', *Science Studies* 1 (1971), 67–83.

The Social Relations of Science Movement was a grouping of left-leaning scientists who were noisily campaigning for a larger role of science in British society.[65] By 1937 they had manifested themselves through booklets such as *The Frustration of Science* (1935), via the editorials of *Nature* and through the writings of the socialist activists and science journalists J.G. Crowther and Ritchie Calder. The core of the movement's message was that science had a great, unused potential for solving society's problems which was far superior to the cumbersome and corrupted ways of existing politics. Whether dealing with issues of poverty, nutrition, housing or the planning of industrial production, scientific methods could make objective analyses and produce solutions that were rational, efficient and fair – not driven by the private interests and partisan battles that characterised the current political system. Some of the movement's proponents located the core of the existing malfunctioning in capitalism, others in parliamentary democracy, but all were united in a preference for scientistic alternatives. One of the movement's allies, H.G. Wells, had long predicted the coming of such scientific forms of government, which, he felt, would also automatically be global, as national differences merely hindered efficiency and had no rational basis in the first place.[66]

Although some of its participants adhered to the Labour Party or the Communist Party of Great Britain, the Social Relations of Science Movement had not grown out of any political organisation.[67] Most members saw its general socialist orientation not as coming from outside but as intrinsic to science itself. Bernal was most outspoken on this point. For him, the ways in which science increasingly shaped society and the historical process leading to a socialist order were so similar as to become essentially identical. He elaborated this identity in his book *The Social Function of Science* (1939) where he concluded, on the last page: 'In its endeavour, science is communism'.[68] Certainly not all of his fellows subscribed to this formulation, but most did see a great deal of congruity between the promises of science and the ideals of socialism. Under socialist conditions, things would be handled scientifically, and a scientifically organised

---

65 William McGucken, *Scientists, Society, and State: The Social Relations of Science Movement in Great Britain 1931–1947* (Columbus 1984). The movement's characterisation given here holds only for this particular phase of its development (the late 1930s through the war years), when left-wing radicals were joined by establishment voices such as Gregory's. Previously, their takes on, for instance, capitalism and the Soviet Union had widely diverged. See Gary Werskey, *The Visible College: A Collective Biography of British Scientists and Socialists of the 1930s* (2nd ed.; London, 1988), 234–49.

66 See Chapter 6, this volume. Wells had only lately aligned himself with the movement's radicals.

67 Werskey, *Visible College* (n. 65), 149–53.

68 J.D. Bernal, *The Social Function of Science* (London, 1939), 415.

society would de facto be socialist. They also attributed similar values to science and socialism: the uplift of humanity, the (ultimate) transcendence of class distinction, anti-militarism, internationalism.[69]

That the Social Relations of Science Movement would support the Dutch 1937 proposal was far from coincidental. Kruyt and Burgers had been corresponding with some of its members, and the entire strategy at the ICSU Assembly (including the distribution of the Memorandum and the *Nature* articles) resulted from deliberations with Gregory.[70] In fact, the proposal itself was largely inspired by the movement. Kruyt and Burgers had followed the British publications closely and in the Memorandum they referred to several *Nature* editorials as well as to works by H.G. Wells, of which Burgers was an avid reader.[71] Moreover, the Memorandum mentioned these writings as examples of the kinds of views on the social role of science that the proposed committee was supposed to collect and publish. Hence what the Dutch scientists sought to do was in fact to advance and disseminate the calls for action that had hitherto come out of the British Social Relations of Science Movement. That was the ultimate purpose of the 1937 proposal.

The Dutch initiatives of the 1930s therefore cannot be understood solely within the Dutch tradition of international mediation. They were also closely related to the Social Relations of Science Movement. This British background helps to explain a number of their outstanding features. First of all, it elucidates why Kruyt and Burgers did not mind omitting explicitly political passages from their 1937 proposal and how the CSSR could follow the original plan after the ICSU general assembly had confined it 'strictly to scientific activity'. The reason was that they had the same conception of the relation of science to politics as the Social Relations of Science Movement. Like their British colleagues, Kruyt and Burgers thought of science as inherently socialist in its method, aim and values. The response to propaganda, nationalism, militarism and so forth was not an added political element, but intrinsic to science itself. Therefore they had no trouble scrapping explicit statements of what was implicitly present anyway. And they could maintain that the CSSR *was* doing strictly scientific work – the political positions that the committee collected were scientific viewpoints at the same time.

---

69 Partly this reflects the flexibility of the term 'socialism' in this period. Most of the movement's members saw themselves as socialists, while few formally aligned themselves with communism.

70 Kruyt and Burgers to Van der Hoeve, 21 January 1937, SU.

71 'Memorandum' (n. 38), 5; Burgers, autobiographical sketch (n. 49), 24–6. One of Wells's writings that the Memorandum referred to was his proposal for a World Encyclopaedia. See also Chapter 6, this volume.

What the connection to the Social Relations of Science Movement also explains is the changed conception of science underlying the Dutchmen's internationalism. The 1934 and 1937 proposals echoed the British writings in their instrumentalism: science was primarily a method, an efficient instrument for solving society's problems and alleviating social needs. The previous notion of science as a cultural good, a proud production of the nation, like art, philosophy, etcetera, had vanished into the background. Relatedly, and finally, British links help to elucidate why the nation no longer took centre stage in the 1930s initiatives. In their projection of science as a counterforce to nationalism, Kruyt and Burgers aligned themselves with H.G. Wells's repeated assertion that in a scientific society, national differences would be obsolete. The scientific way of handling things implied a retreat of the nation state – in Wells's writings as well as in the proposal texts. The proposals also stopped emphasising and legitimating the Dutch national calling as mediators. In fact, the initiatives were no longer very Dutch at all – Kruyt and Burgers had made great efforts to have them backed by an international network even before they were presented to the general assembly. What remained of the older internationalist tradition was the fact that the initiative came from the Amsterdam Academy. But much of its conceptions and strategy were related to the Social Relations of Science Movement.

Unified unifications

While the Dutch internationalisers collaborated with the British Social Relations of Science Movement, neither of them had significant interactions with Otto Neurath's Unity of Science Movement – even though Neurath lived in The Hague from 1934 to 1940 and in Oxford until his death in 1945.[72] Still there are some striking similarities between the two endeavours. The implied socialism, the anti-nationalism and the view of science as an alternative to existing ways of government that we now know from the Anglo-Dutch cooperation were also strongly present in Neurath's movement. In either case, internationalism did not stand alone but was part of a larger socialist-scientific programme. Or – formulated in terms of unity of science pursuits – each sought to unify nations as well as scientific and societal objectives. Parallelisms like these are elusive phenomena, however, and it would be beyond the scope of this chapter to try to explain the similarities between the Anglo-Dutch and Viennese projects. But what we can do is follow some of its implications that the resemblance suggests with respect to the Dutch and British activities.

If the Dutch scientists' internationalism was anything like that of Neurath's Unity of Science Movement, it would be surprising if it were an isolated affair,

72 For one important exception, see Chapter 8, this volume.

cut loose from other political concerns. And indeed if we look at Kruyt and Burgers's broader activities we see several related socialist-scientific projects. Kruyt, for one, agitated throughout the 1930s for a much more extensive use of science in society. He helped to create a national organisation for applied scientific research, literally called Applied Scientific Research, which aimed to place virtually all economic sectors (agriculture, industry, fishery, but also healthcare, nutrition and so on) on a scientific basis. The only resistance to such scientification, he claimed, came from the political establishment, 'an alien element' in the chain linking science and society.[73] Similarly, Kruyt tried to align scientific education to social needs. In a controversial speech in 1931 he denounced the Dutch universities for having become conservative institutions of isolated academic learning while they should be the engines of modern society, providing all kinds of professions with a scientific training. Students especially appreciated this message, and Kruyt got involved in a series of initiatives to reform higher education.[74] In 1946 he left academia to preside over the Applied Scientific Research organisation.

Burgers's activity was not limited to internationalism either, but his engagement took a more philosophical route.[75] For him, the relation of science to socialism became a subject of further inquiry. How did we know scientific approaches would build a better (socialist) world if science itself was value-neutral? How to make sure that amoral knowledge was used for good and not for bad? Here, it was the books by A.N. Whitehead and the new quantum mechanics that inspired Burgers to a solution: an idea of a value-*laden* science with ethical choices built into its practice. Such a 'synthesis' of research and morality could guarantee that a scientific society was also a socialist one.[76]

---

[73] Somsen, *Wetenschappelijk Onderzoek* (n. 46), 197–206: 204. Geert J. Somsen, 'Selling Science: Dutch Debates on the Industrial Significance of University Chemistry, 1900–1932' in A.S. Travis et al. (eds), *Determinants in the Evolution of the European Chemical Industry, 1900–1939: New Technologies, Political Frameworks, Markets, Companies* (Dordrecht, 1998), 143–68: 156–66.

[74] Somsen, *Wetenschappelijk Onderzoek* (n. 46), 191–7.

[75] G. Alberts, *Jaren van Berekening. Toepassingsgerichte Initiatieven in de Nederlandse Wiskundebeoefening 1945–1960* (Amsterdam, 1998), 120–24; Alberts, 'Connecting Socialism and Mathematics' (n. 49). Geert J. Somsen, 'Value-Laden Science: Jan Burgers and Scientific Politics in the Netherlands', *Minerva* 46 (2008), 231–45.

[76] J.M. Burgers, 'Synthetische Wijsbegeerte' and 'Gemeenschappen (Leer der)' in H.J. Jordan et al. (eds), *Encyclopaedisch Handboek van het Moderne Denken* 2nd ed. (Arnhem, 1942), 726–9, 280–85; idem, 'Trekken van de Moderne Westerse Wetenschap', *Mededeelingen van de Nederlandsche Akademie van Wetenschappen, Afdeeling Letterkunde* (Amsterdam, 1944), 9–15.

Burgers expressed these ideas in philosophical writings, but he also tried to implement them in political activities.[77] Shortly after the end of the Second World War he established a 'Study Centre for Social Issues' in order to serve society through 'scientific-synthetic' activities. Under the banner of the Study Centre, various scientists started to work out scientific solutions for pressing social problems, and cooperation was started with existing planning institutions and with several ministers of the first post-war administration, which was highly technocratic. Already before the liberation, Burgers and a few others had drafted a general structure for a scientifically planned society, and at some point it looked as if his Study Centre would be asked to turn this blueprint into a new Dutch constitution. This plan fell through as the technocrats lost the 1946 elections, but the episode illustrates the extent of Burgers's scientific socialism.[78]

With respect to the British Social Relations of Science Movement, the parallel with Neurath's programme suggests that it should also incorporate internationalist endeavours. Here the picture is more scattered, however. For a long time, many in the movement acknowledged but few stressed the internationalising implications of a greater role of science in society – save H.G. Wells. Most campaigning was focused on problems within Britain.[79] This only started to change after 1938, when the British Association for the Advancement of Science set up a new Division for the Social and International Relations of Science (DSIRS). The aim of the DSIRS was to hook up with similar movements abroad, and during the Second World War it developed into a major carrier of socialist scientific internationalism (see the epilogue below). It started out, however, as a national subsection of the CSSR, and hence, it originated partly from the 1937 ICSU resolution. Therefore the Dutch initiatives helped to add internationalism to the agenda of the Social Relations of Science Movement, albeit not just any internationalism. As this chapter has shown, different understandings of the international character of science dominated before the 1930s. It was a new and distinctly socialist internationalism that informed unity of science pursuits in the interwar period.

---

77 Burgers ended up publishing two philosophical monographs, the second a reworked translation of the first: *Ervaring en Conceptie* (Arnhem, 1956) and *Experience and Conceptual Activity: A Philosophical Essay Based upon the Writings of A.N. Whitehead* (Cambridge, MA, 1965).

78 Geert Somsen, 'Waardevolle Wetenschap. Bespiegelingen over Natuurwetenschap, Moraal en Samenleving in de Aanloop naar de Doorbraak-Beweging', in Geert Somsen (ed.), *De Doorbraak van de Experts. Wetenschap en Maatschappelijke vernieuwing Rond 1945* (Rotterdam, 2001), 19–36.

79 For earlier international contacts, see Patrick Petitjean, 'Needham, Anglo-French Civilities and Ecumenical Science', in S. Irfan Habib and Dhruv Raina (eds), *Situating the History of Science: Dialogues with Joseph Needham* (New Delhi, 1999), 152–97.

Epilogue

What happened to this socialist scientific internationalism after the late 1930s? For the Dutch scientists, the Second World War seriously hindered international activity, and it was virtually impossible to continue their endeavours under Nazi occupation. Burgers could not finish the report on the social relations of science, and the CSSR ceased to function in general. In Britain, however, the DSIRS became very active during the war. Presided by Gregory, it organised a series of events that continued to promote socialist-scientific views of reshaping society, nationally as well as internationally. Of these, the conference Science and World Order of 1941 was probably the largest and most widely publicised manifestation.[80] Gathered in the Royal Institution, scientists from 29 countries met with state officials to discuss how a new society could be built up after the war with the aid of scientific methods and in keeping with scientific values. There were sessions on 'Science and World Planning', on education, on government – and on internationalism. Once again, international cooperation was presented as an essential feature of a scientific society and nationalism condemned as its antithesis. This time, nationalism was also associated with Aryan physics and Nazi ideology – a connection that compromised it even further.[81]

After the war internationalist activity quickly resumed, also outside of Britain. In October 1945 the Amsterdam Academy submitted another resolution, and in December the ICSU, now under Kruyt's presidency, commissioned a new CSSR report. The main concerns had changed: *the* issue was now atomic energy and the secrecy restrictions surrounding the new know-how. But the solutions were still sought along socialist-internationalist lines. According to the Academy proposal, the only way out for the world was by maintaining scientific exchanges, seeking international unity and giving greater political power to 'the world of science'.[82] The CSSR voiced similar views. In a set of resolutions, Burgers identified military influence and private enterprise as the chief sources of the current dangers, claiming that science demanded 'world citizenship' and the 'education of . . . governments in the aims, methods and spirit of scientific research'.[83]

---

80 *The Advancement of Science* 2 (1942), 1–120. Paul Rotha made a film of the conference – see Chapter 8, this volume.

81 J.G. Crowther, O.J.R. Howarth and D.P. Riley, *Science and World Order* (Harmondsworth and New York, 1942), 129–38.

82 *Report of the Committee on Science and Its Social Relations (C.S.S.R.) on the Public Attitude of Scientific Bodies with regard to the Problems Raised by Recent Applications of Science* (London, 1946), 9.

83 'Communication from the C.S.S.R.' [summer 1948], typescript, IUPAC files (hereafter 'IUPAC'), Chemical Heritage Foundation, Philadelphia. See also F.J.M. Stratton

But while socialist internationalism returned quickly, so did the resistance against it. Some of this was similar to the objections raised in 1937 that the political ambitions of the left-wing scientists were too large and explicit. At the 1946 ICSU assembly, Royal Society President Robert Robinson criticised the 'political paramountcy' of the newest plans, while questioning that 'loyalty to science must [necessarily] transcend loyalty to King and country'.[84] An opinion poll among 70 scientists which the CSSR held in 1948 showed that actually very few of the respondents wanted direct political influence and some were even sceptical about the effect of science on international politics.[85] As before, the socialist view turned out to be prominent, but certainly not shared by all.

New objections appeared as well. In Britain, Bernal and others' support of the Soviet Union (as the most scientific society existent) was increasingly criticised, and similar voices were raised in the CSSR and in Dutch circles. The Amsterdam physicist Jacob Clay, for instance, noted that the international cooperation of scientists was actually obstructed by communist regimes and demanded that his left-wing colleagues speak to those governments.[86] Dutch scientists of a younger generation also began to question the leftist and anti-militarist activism, if only because it might turn off the government and decrease scientists' influence.[87] From another angle, the CSSR observed that security or loyalty investigations started to scare young scientists out of their political engagement, 'since better positions will generally be available to those who "keep quiet" than to those who do not.'[88] It would exceed the scope of this chapter to examine the exact effect of these anti-communist developments (see Reisch in this volume for a study that does), but it is clear that the socialist conception of science and internationalism was facing hard times. Burgers's reports for the CSSR were increasingly amended and stripped from their

(ed.), *The Fifth General Assembly of the International Council of Scientific Unions held at Copenhagen, September 14th to 16th 1949. Reports of Proceedings* (Cambridge, 1950), 82 and 86–7.

84 F.J.M. Stratton (ed.), *The Fourth General Assembly of the International Council of Scientific Unions held at London, July 22nd to 24th 1946. Reports of Proceedings* (Cambridge, 1946), 2.

85 F.J.M. Stratton (ed.), *Report of the Executive Committee 1948* (Cambridge, 1949), 63–86, esp. 68, 74–5.

86 Ibid., 82. Clay was the CSSR's correspondent for the Netherlands.

87 These sentiments were voiced in the Dutch *Verbond van Wetenschappelijke Onderzoekers* (Association of Scientific Workers), established by a large group of young Dutch scientists in 1946. See Leo Molenaar, *'Wij Kunnen het Niet Langer aan de Politici Overlaten . . .' De Geschiedenis van het Verbond van Wetenschappelijke Onderzoekers (VWO) 1946–1980* (Delft, 1994), 94–9.

88 Stratton, *Fifth General Assembly* (n. 83), 83.

anti-militarist statements.[89] And the Amsterdam Academy 1945 resolution received very little support. Of all major academies it had been sent to, only those of Belgium, Sweden and France signed their approval, Britain and the US being conspicuously absent.[90] Like many other unifying pursuits discussed in this volume, socialist internationalism started to lose its support and its prominence in the wake of the Cold War.

Acknowledgements

Part of this work was made possible by a Chemical Heritage Foundation Travel Grant. Anna-K. Mayer, Margaret Meredith and Harmke Kamminga gave help and comments, for which I am and remain grateful.

89 This becomes clear from a comparison of drafts to published reports, for example J.M. Burgers, 'Confidential Draft Report, 25 June 1946' (IUPAC) to *Report of the Committee* (n. 82) and 'Communication' (n. 82) to Stratton, *Fifth General Assembly* (n. 83).

90 *Report of the Committee* (n. 82), 9–10.

Chapter 6
Government as scientific process in H.G. Wells's world state

Rein de Wilde and Geert Somsen

Wells's world state

'Unity' and 'science' were the two basic concepts, and strongly interconnected ones, in a project that the famous writer H.G. Wells pursued during the interwar period: the establishment of a world state. In numerous publications, Wells tried to move his fellow citizens to overcome national boundaries and unite into a global society. The difference with rival utopias of the period, like those promised by communism and Christianity, was that this form of paradise was not to be achieved by revolution or salvation, but primarily by science. In Wells's vision of the rise and flourishing of the world state, scientific knowledge and scientific procedures played major parts. Here we examine why Wells made such a strong association between science and the political unification of the world.

During the first half of the twentieth century, Herbert George Wells (1866–1946) was one of the best known British novelists and political writers in the Western world. Before the First World War, he wrote dozens of scientific romances, sociological novels and socialist tracts illuminating the breakdown of the Victorian ancien régime and preaching the gospel of science, progress and socialism. These found a ready audience in Britain and abroad. Wells rejoiced in the steady advance of human knowledge and its applications, and gratefully held them responsible for the decay of the old establishment.

Wells was not alone in opposing contemporary society, but unlike nearly all his fellow critics, he included the nation state in his indictment of the European order. As early as 1902, in a series of *Anticipations*, he predicted a colossal world war among the nations. This war, he hoped, would end in the seizure of real power throughout the world by a new breed of men conscious of the fatal futility of nationalism and tough-mindedly prepared to reorganise the world in a 'larger synthesis'.[1] These 'new republicans' were the first draft, as it were, of the later 'samurai', the ruling class in *A Modern Utopia* of 1905, which was

[1] H.G. Wells, *Anticipations of the Reaction of Mechanical and Scientific Progress upon Human Life and Thought* (Mineola, 1999 [1902]), 138–56.

Figure 6.1 Cartoon of H.G. Wells by David Low.
Source: Reproduced with permission, Solo Syndication, London.

Wells's thought experiment on the future. Loosely based on the Japanese elite fighters, Wells called these samurai a 'voluntary nobility', 'men and women of a distinctive costume and bearing'.[2]

In his introduction to *A Modern Utopia* Wells told his readers that no less than a planet would serve the purpose of a modern utopia:

> Time was when a mountain valley or an island seemed to promise sufficient isolation for a polity to maintain itself intact from outward force; the Republic of Plato stood armed ready for defensive war, and the New Atlantis and the Utopia of More in theory, like China and Japan through many centuries of effectual practice, held themselves isolated from intruders.... But the whole trend of modern thought is against the permanence of any such enclosures.... Now, perhaps you might still guard a rocky coast or a narrow pass; but what of that near to-morrow when the flying machine soars overhead, free to descend at this point or that? A state powerful enough to keep isolated under modern conditions would be powerful enough to rule the world.... World-state, therefore, it must be.[3]

According to Roslynn Haynes, of the utopian characteristics that Wells advocated, this stress on globality was perhaps his most important contribution

2 H.G. Wells, *A Modern Utopia* (Lincoln, 1967 [1905]), 259.

3 Ibid., 11–12.

to the history of utopian thought.[4] Warren Wagar even calls Wells 'the only outstanding prophet of world order in the first third of the [twentieth] century'.[5] The kind of government Wells envisaged in his future state was technocratic and socialist. It was to be ruled by an elite of science-minded 'functional men' who would seize power during a crisis and retain it through their efficiency and general ability. Wells believed that mental reform must precede political reconstruction, and hence the ruling samurai represent our 'best selves'. Throughout his career, Wells believed that an extensive eugenics programme and a uniform educational system would prevent the survival of inferior individuals after only a few generations. The samurai class would then encompass virtually all members of the world state, as it does in his book *Men Like Gods* of 1923.[6]

In the Wellsian future, unification at the state level went hand in hand with a more comprehensive cultural unification. Not only should all nation states merge into one world state, systems of thought that challenged the ideal of a unified world must disappear, too. Wells heavily attacked Zionism, for instance, in his essay *The Fate of Homo Sapiens*, written in 1939.[7] In his futuristic novel *The Shape of Things to Come* of 1933, he had already scolded orthodox Jews for not being in tune with 'the logic of history'. The Zionist endeavour, he wrote, was the embodiment of 'their traditional wilful separation from the main body of mankind'. Palestine was a misguided 'dream of a fantastic independent state all of their own'. But while Wells was writing these things at the time that the Nazis had started persecuting Jews in Germany, his objections were different in that they were not couched in terms of race. He had long rejected the notion of human races as lacking scientific basis and believed that the interbreeding of populations would continue to increase. His intolerance was with cultural separatism. And it is that aspect of the Jewish people that he predicted would be made to disappear the future state:

> The Jews were not suppressed; there was no extermination; . . . yet they were educated out of their oddity and racial egotism in little more than three

[4] Roslynn D. Haynes, *H.G. Wells, Discoverer of the Future: The Influence of Science on his Thought* (New York, 1980), 92.

[5] Warren W. Wagar, *The City of Man: Prophecies of a World Civilization in Twentieth-Century Thought* (Boston, 1963), 59. See also John S. Partington, *Building Cosmopolis: The Political Thought of H.G. Wells* (Alderhot, 2003).

[6] H.G. Wells, *Men Like Gods* (New York, 1923). On Wells's commitment to eugenics, see Michael Coren, *The Invisible Man: The Life and Liberties of H.G. Wells* (London, 1993).

[7] H.G. Wells, *The Fate of Homo Sapiens: An Unemotional Statement of the Things that are Happening to Him Now, and of the Immediate Possibilities Confronting Him* (London, 1939).

> generations.... The world is as full as ever it was of men and women of Semitic origin, but they belong no more to 'Israel'.[8]

Wells did not single out Zionism and Jewish orthodoxy as enemies of the world state either. Sooner or later, he claimed, mankind must come to one universal peace, which will take the form of a world government. Given this future of the planet, he praised the Catholic Church as 'the first clearly conscious attempt to provide such a government in the world'.[9] But after a lengthy review of its policies through the ages, Wells's conclusion was equally relentless:

> Before mankind gets rid of it, the Papacy may be drowning our hopes for the coming generation in a welter of blood [by silently endorsing Hitler] ... and it will not add an inch to his stature nor alter the fact that the Pope, any Pope, is necessarily an ill-educated and foolish obstacle ... in the way to a better order in the world.[10]

Clearly, Wells's idea of a world state is at odds with today's dominant values. Many who are now left-leaning and academically educated, like Wells was, don't cherish global uniformity but cultural diversity. And those who criticise multiculturalism often have an equal aversion to supranational impositions which aim to standardise customs, language and taste. But to Wells, unification meant conformity to one standard, and as we aim to show, that standard was rooted in the style and thought of the British educated middle class. Liberty and individuality were limited in the sense that nobody was allowed to question the basic assumption that world unity is the supreme human value. Wells is often praised for the fact that he left room for change in his utopian schemes, yet if we look more closely, change was only permitted as long as society and its members moved in the 'right' direction of unification.

The way to the world

Wells rose to fame and fortune from a lower-middle-class background. His handicapped father kept a dwindling shop, and his mother worked as a maid. Wells was apprenticed as a draper, but he managed to escape this future by being admitted, first, to a grammar school and then to the Normal School of Science

---

8 H.G. Wells, *The Shape of Things to Come: The Ultimate Revolution* (London, 1993 [1933]), 373. The idea of extermination here did not come from the Nazis but from Wells's own suggestion, in 1902, that in the future people out of tune with scientific modernity would be euthanized. See Wells, *Anticipations* (n. 1), 316–17. In subsequent publications Wells distanced himself from this proposal.

9 H.G. Wells, *Crux Ansata: An Indictment of the Roman Catholic Church* (Harmondsworth, 1943), 10.

10 Ibid., 96.

in South Kensington, London. The latter had a major formative influence on him. Not only did he strike up lifelong friendships there with people of a similar background who continued to share the same orientation and often similar pursuits – most notably Richard A. Gregory, the later editor of *Nature* – he also underwent a particular kind of schooling that would foster his convictions of the power and importance of science, facts and education. The Normal School of Science was a teacher training institution, mainly geared toward supplying staff for the many new schools that sprang up after the 1870 Education Act. It was neither a place of higher learning nor a technical school, but deliberately geared toward practice as well as theory: knowledge was not just conceived within the realms of the mind, but also as a tool, available to man for understanding and improving his place in the world. Although Wells would not end up in the teaching profession proper, he continued to see himself as an educator, and took the Normal School's orientations into his fictional and non-fictional writing.[11]

The teacher in South Kensington with the biggest impact on Wells ('the greatest man I was ever likely to meet') was Thomas Henry Huxley, the biologist and ardent advocate of professional secular science, who would become famous as 'Darwin's bulldog'.[12] One of Huxley's main arguments with respect to evolution was that its optimistic interpretation as 'progressive development' was based on a profound misconception. There was no discernible purpose in natural selection and nothing to justify a belief in automatic progress. The collective human will, however, was not necessarily subject to blind selective forces. It was an agency, peculiar to mankind, that stood outside and against the cosmic process of evolution, challenging it with a specifically human purpose. This did open up the possibility of progress.[13] Wells followed Huxley on both counts (the absence of automatic progress and man's ability to control his own evolution), as would Huxley's grandson Julian who later elaborated these ideas in his philosophy of 'evolutionary humanism'.[14] Wells got to know Julian Huxley very well as they worked together on the biology outline, *The Science of Life*, in the late 1920s.[15] But the initial impact had come from the grandfather.

---

11 David C. Smith, *H.G. Wells: Desperately Mortal: A Biography* (New Haven, 1986), 3–20.

12 Ibid., 11.

13 See especially Thomas Huxley's Romanes lecture 'Evolution and Ethics', delivered in Oxford in 1893 and published, with introductory prolegomena, in T.H. Huxley, *Evolution and Ethics* (London, 1894).

14 Warren W. Wagar, *H.G. Wells and the World State* (New Haven, 1961), 63–5. Krishan Kumar, *Utopia and Anti-Utopia in Modern Times* (Oxford, 1987), 177. On Julian Huxley, see Chapter 3, this volume.

15 H.G. Wells, J.S. Huxley and G.P. Wells, *The Science of Life*, 3 vols (London, 1929–30).

Thomas Huxley's influence also makes understandable that Wells was never a staunch optimist: progress was not guaranteed by natural forces, but depended always on the good will of mankind. Several times in his life, Wells lost confidence in the will and ability of the intellectual elite of his day and, as a consequence, his optimism switched into a pessimistic mood. Likewise, his agreement with Thomas Huxley that 'nothing comes of itself except weeds and confusion' prevented him from presenting science and technology as progressive agents in their own right.[16] The development of science and technology did increase the *need* for a world state, because it made wars more and more devastating. At the same time, science and technology created the possibility of a world state by providing the means to build its infrastructure. Communication technologies like the telegraph and the telephone, and transport technologies like the railways, the motorcar and the 'flying machine' were responsible for what Wells aptly called 'the abolition of distance'.[17] Yet science and technology alone would never *automatically* produce a world state: without the work of a forward-looking human avant-garde, history would never evolve in the right direction. Throughout his life, Wells looked for, or tried to organise, such an 'open conspiracy' of progressive and reasonable people leading the way.

One of the groups he hoped might perform such a role was the Fabian Society, which he joined in 1903, on the invitation of its leaders Sidney and Beatrice Webb. Wells strongly sympathised with the society's non-revolutionary form of socialism, its preference for scientific approaches to social reform and its emphasis on education. He did feel, however, as did some of the other Fabians, that the society should be much more active and outspoken and urgently needed rejuvenation. To this end, Wells started to elaborate ambitious plans, which he laid down in a new 'Basis', that he hoped (in vain) would become the new Fabian charter. Later, in his *Experiment in Autobiography* of 1934, Wells recalled how he wanted to revolutionise the Fabian strategy from top to bottom, to bring the Society into politics and education, to increase its membership from 700 to 10,000, to convert it, in effect,

> into the beginnings of an order, akin to [the] samurai in *A Modern Utopia*. . . . I envisaged that reconditioned Fabian Society as becoming, by means of vigorous propaganda, mainly carried on by young people, the directive element of a reorganized socialist party. We would attack the coming generation at the high school, teaching college and university stage, and our organization would quicken into a constructive social stratum.[18]

---

16 Quoted in Wagar, *Wells and the World State* (n. 14), 80.

17 Wells, *Shape of Things* (n. 8), 12.

18 Quoted in Wagar, *Wells and the World State* (n. 14), 26.

Although several 'Samurai societies' were indeed being formed in Britain around the time that Wells wrote his plans, leading Fabians like George Bernard Shaw were not so ready to give up their autonomy and be recruited into Wells's army of world state warriors. They opposed and obstructed his plans and isolated the author, and within a few years Wells left the Fabian Society bitter and disappointed.[19]

While the Fabian Society was clearly not to be Wells's political home, it is quite difficult to point down what was. He was a self-declared socialist, but his party affiliations seem to have been weak and changing. In the early 1920s, he twice ran for election as Labour MP, without success and without this strengthening his ties to the Labour Party.[20] The Communist Party of Great Britain did not appeal to him either, and he would always retain some reservations toward communist scientists, like J.D. Bernal and J.B.S. Haldane, even though they shared his ideal of a scientifically run society. Obviously, Wells was anything but apolitical, but his politics superseded the level of national parties, preferring the bigger picture of global developments. At that level he did directly engage himself, not only through his writings, but also by conferring, many times in the course of his life, with world leaders from Theodore Roosevelt to Lenin and from Churchill to Stalin.

Throughout his extremely versatile active life, Wells operated on many fronts. Still, the circles closest to him were the world of science, education and publication in Britain. Although he was never a practising scientist himself, he was well acquainted with many of them and with their work. On the front of education, he and his friend Gregory were active in the educational section 'L' of the British Association for the Advancement of Science (BAAS) and in several science-promoting committees of the British Science Guild. After its establishment in 1938, Wells also became very active in the BAAS's Division for the Social and International Relations of Science and in the conferences it organised during wartime. Just as England was the home from which he viewed the whole world, so scientific education determined his outlook on all spheres of life.

The necessity for a world state became even more urgent for Wells after the Great War. Its massive atrocities and relentless continuation convinced him that the real fault was less with any particular country than with the institution of the nation state. Wells 'replayed' the process of acquiring this insight in his wartime novel *Mr. Britling Sees It Through*, where an Englishman and a German correspond over the frontlines and learn to recognise how similar and unnecessary their sufferings really are.[21] Wells hoped that after the war its deeper

---

19 Smith, *H.G. Wells* (n. 11), 105–11. On the 'Samurai societies' (set up mainly to discuss Wells's work), see 101.

20 Ibid., 275–8.

21 H.G. Wells, *Mr. Britling Sees It Through* (London, 1916).

causes would be dealt with and that some sort of supranational government would be established. He actively pursued this goal, with some kindred spirits, in a League of Free Nations Society, publishing, among other things, the pamphlet *The Idea of a League of Nations*. This idea, however, was a much stronger and more comprehensive central body than the League that would eventually arise from the Versailles meetings. Wells was disappointed with this outcome, but his conviction of the necessity of a world state had only grown in the process.[22]

Always ambivalent about the prospects for progress, later in life Wells became disillusioned by the rise of fascism. Yet he threw himself into positive causes, such as the Social Relations of Science Movement and especially his campaign for a 'New Declaration of the Rights of Man'. Throughout, he never abandoned his ideal of the world state, a theme that is present in almost all his works, fiction and non-fiction alike. However, surveying his impressive oeuvre, the question that remains unanswered is why this idea of a world state was so central to Wells. In the literature on Wells, most scholars are satisfied with his own twofold answer: first, that a world state is necessary because nation states make war, so world peace is not possible without their abolition, and second, that global 'overpopulation' was imminent and only a world government could control human breeding.[23]

Neither of these reasons is very convincing. Wells's fears about overpopulation sat uncomfortably with his general beliefs in the powers of science and technology to transform the world. They were also quite unusual. Especially in the 1930s, the technocratically inclined in Britain rather held that science and technology were capable of producing a 'world of plenty' and that any possible lack of food in the future could be countered by measures such as improvements in agricultural techniques.[24] Secondly, why should war and aggression cease to exist in a world state? Already in 1905, one of Wells's closest friends *and* chief adversary, the conservative writer G.K. Chesterton, denied strongly that it would:

> It is impossible to prevent a possible conflict of civilisations, because it is impossible to prevent a possible conflict between ideals. If there were no longer our modern strife between nations, there would only be a strife between Utopias. For the highest thing does not tend to union only; the highest thing tends also to differentiation.[25]

---

22 Smith, *H.G. Wells* (n. 11), 241–2.

23 See e.g. Kumar, *Utopia and Anti-Utopia* (n. 14), P. Parrinder, *Shadows of the Future: H.G. Wells, Science Fiction and Prophecy* (Syracuse, 1995) and Wagar, *Wells and the World State* (n. 14).

24 See Chapter 8, this volume.

25 G.K. Chesterton, 'Mr Wells and the Giants', in idem, *Heretics* (London, 1905), 68–91: 80.

Chesterton's criticism laid bare one of Wells's most basic assumptions: that 'the highest thing' *does* 'tend to union'. To Wells this highest thing was science, and it was unified at heart: there was not more than one kind of scientific truth or rationality. Wells also believed that no limits could or should be set to science's sphere of influence. This 'expansionist' view in combination with the idea of unified science entailed that there would no longer be a diversity of nations and cultures – and no strife between them.[26] This, we suggest, explains to a large extent Wells's lifelong conviction that a modern utopia can be nothing else than a globally unified society. In effect, Wells's world state was his image of science writ large.

Inner workings

Now how exactly did Wells think his world state would operate and be organised? In *The Open Conspiracy* of 1933, he warned his readers that it would be fundamentally different from 'the patch-work of today'. First, the world state should *not* be conceived as a nation state on a global scale. 'This point is not yet generally realised. It is too often assumed that the world commonweal will be, as it were, just the one heir and survivor of existing states, and that it will be a sort of megatherium of the same form and anatomy as its predecessors'.[27] According to Wells, that is a mistake:

> Existing states are primarily militant states, and a world state cannot be militant. There will be little need for president or king to lead the marshalled hosts of humanity, for where there is no war there is no need of any leader to lead hosts anywhere, and in a polyglot world a parliament of mankind or any sort of council that meets and talks is an inconceivable instrument of government. The voice will cease to be a suitable vehicle. World government [will be] like scientific process.[28]

All the questionable claims Wells made in this passage are placed in perspective by his final analogy: world government as scientific process. This analogy cannot be taken seriously enough. Because, according to Wells, science proceeded neither by autocratic nor by democratic procedures, it was inconceivable to him that his ideal state could be organised in either of these ways. Autocratic rule would be too close to the ways of the traditional nation states or their feudal predecessors. But democratic forms of government met with Wells's reservations too. In several papers from the 1920s through the 1940s, he explained his 'doubts

---

26 The analytic category of 'expansionism' is borrowed from Loren Graham, *Between Science and Values* (New York, 1981).

27 H.G. Wells, *The Open Conspiracy and Other Writings* (London, 1933), 30.

28 Ibid., 30–31.

on democracy'. They were twofold, pertaining both to the nature of the ruling class of elected politicians and to the capacities of their public, the voters.[29]

The way the democracies were organised meant that elected politicians did not necessarily have to be intellectually and morally responsible people. They were trained as politicians and generally concerned with their own re-election and not with the real needs of the public. If they had been, the leaders of the British, French and American democracies certainly should have been more effective in preventing yet another world war from happening. But they had failed on this count and therefore, according to Wells, '[i]t is an open question whether they are much more responsive to popular feeling than the dictators we denounce so unreservedly as the antithesis of democracy'.[30]

The trouble with the voting public was that it was fed heaps of information and misleading propaganda. Being insufficiently educated, voters had no way of knowing which vote would really further their interests:

> It appears that we in our muddled multitudinousness are being called upon to make decisions. This immense tangled affair, we gather, is our affair. In various rather obscure ways we have been made responsible for it. We have to vote. But how can we vote when we do not get the hang of it? Has anyone got the hang of it? Are there people anywhere in our world today who have a really comprehensive vision of the economic world process as one whole? Apparently not. And yet we ordinary people have it thrust upon us, that whatever control can be exercised over this immense complex tumult of world change, must be exercised through our voting and our assent.[31]

The way out, and the way to go for the world state, was government through scientific process. But what did Wells mean by that? It did not imply, he was quick to point out, that scientists themselves should rule the world. 'Some favour the idea of a gradual supersession of the political forms and methods of mass democracy by some sort of élite, in which the man of science and the technician will play a dominating part'.[32] This was a bad idea, according to Wells, because scientists as scientists are good specialists at best, but taken together they are no more than 'a miscellaneous assembly'. Therefore, it is '*science* and not *men of science* that we want to enlighten and animate our politics and rule the world'.[33]

---

29 'Doubts on democracy' was the title of a lecture Wells gave at the Sorbonne, Paris, in 1927. See also Charlotte Thomsen, 'H.G. Wells and the World State' (MA thesis, Maastricht University, 2000), 30–32.

30 Wells, *Outlook for Homo Sapien* (n. 7), 234.

31 H.G. Wells, *The Outline of Man's Work and Wealth* (Garden City, 1936), 13–14.

32 H.G. Wells, 'World Encyclopaedia', in idem, *World Brain* (London, 1938), 1–25: 9.

33 Ibid., 11; emphasis in original.

The specialised knowledge of experts was not enough – what was needed was a synthesis of this knowledge. This had become painfully clear in the aftermath of the Great War, when, according to Wells, all the knowledge and all the directive ideas needed to establish 'a wise and stable settlement of the world's affairs' existed, yet only in bits and fragments. '[P]ractically nothing had been assembled, practically nothing had been thought out, nothing practically had been done to draw that knowledge and those ideas together into a comprehensive conception of the world'.[34] As a consequence, a stable peace was never reached, and it was this experience that made Wells aware how urgent the world's need was for synthesised knowledge.

Immediately after the disappointment of the Versailles negotiations, Wells realised that a grand task lay ahead of him. The fact that leaders had failed to see the true cause of the war (the institution of the nation state) stemmed from their limited knowledge and poor education – an education that, where international relations were concerned, was invariably based on partial and nationalist histories. No wonder that nationalism had prevailed. To remedy this, Wells undertook to write a completely new historical overview, one that was not particular but global in outlook – a history, not of a nation, but of the human species: *The Outline of History*.[35] After this overview, he went on to produce an outline of biological insights, *The Science of Life*, and a unified presentation of the economic and social facts of life, *The Work, Wealth and Happiness of Mankind*.[36] All of these works were the sorts of syntheses that Wells had in mind, providing the reader with the general knowledge and global vision that the world needed. They were widely sold, but not enough, it appeared by the early 1930s, to bring international cooperation much closer. So, together with Gregory, Wells decided it was time for something even bigger: a campaign for a World Encyclopaedia.

Encyclopaedism

Wells's World Encyclopaedia (he sometimes spoke of a world brain or world mind) was to be an all-encompassing, up-to-date, yet concise collection of knowledge. It should consist of 'selections, extracts, quotations, very carefully assembled with the approval of outstanding authorities in each subject, carefully collated and edited and critically presented. It would not be a

---

34 Ibid., 4.

35 H.G. Wells, *The Outline of History*, 2 vols (London, 1919–20), predecessor to many translations and other versions, among which *A Short History of the World* (London, 1922). On the decision to write this book, see Smith, *H.G. Wells* (n. 11), 249.

36 Wells, Huxley and Wells, *Science of Life* (n. 15); H.G. Wells, *The Work, Wealth and Happiness of Mankind* (London, 1932). On this project, see also Chapter 3, this volume.

miscellany, but a concentration, a clarification and a synthesis'.[37] The project of its assemblage should be undertaken by an encyclopaedia organisation, consisting of the best experts in every field, 'a super university' as Wells called it, 'a world of brain'.[38] Not only would the members of this organisation compile knowledge, they would also revise, extend and replace what was already compiled, and do so continuously. The World Encyclopaedia would never be a finished product. Like the world state, it would continue to develop and improve itself.

Now what role would this encyclopaedia have in a future world state? First of all, it would have different uses for different audiences. For the 'ordinary educated citizen' the World Encyclopaedia would be 'a row of volumes in his own home or . . . in a convenient public library'. It would present him with the ruling concepts of social order, outlines and main particulars in all fields of knowledge and, in fields where expert opinion was divided, 'very carefully chosen and correlated statements and arguments'.[39] For educators, the World Encyclopaedia would be a major resource in their teaching activities. In 1937, Wells developed an entirely new school curriculum that stretched from infancy to adulthood and would explain all essential knowledge (Figure 6.2). The World Encyclopaedia was to be a main organising principle in this programme, especially in the more advanced stages.[40]

For the expert, the World Encyclopaedia would be of value because 'it is going to afford him an intelligent statement of what is done by workers parallel with himself'.[41] Thus, it could lift scientists out of their narrow-minded and pedantic specialisation and make them more 'world-conscious'. For universities, the encyclopaedia was a route back to the power they had lost through their conservatism and emphasis on classics and literature.[42] And for future government, finally, the World Encyclopaedia, would present the facts and ideas so necessary for adequate understanding and judicious decision making. 'Ultimately if our dream is realised it *must* exert a very great influence upon everyone who controls administrations, makes wars, directs mass behaviour,

37 Wells, 'World Encyclopaedia' (n. 32), 14.

38 W.H.G. Armytage, *Sir Richard Gregory: His Life and Work* (London, 1957), 151.

39 Wells, 'World Encyclopaedia' (n. 32), 13.

40 Wells ventured these ideas on several occasions. See for example his 1937 presidential address to the Educational Science section of the BAAS: 'The Informative Content of Education', *BAAS Report of the Annual Meeting, 1937 (107th year), Nottingham, September 1–8* (London, 1937), 237–48.

41 Wells, 'World Encyclopaedia' (n. 32), 16.

42 Ibid., 12.

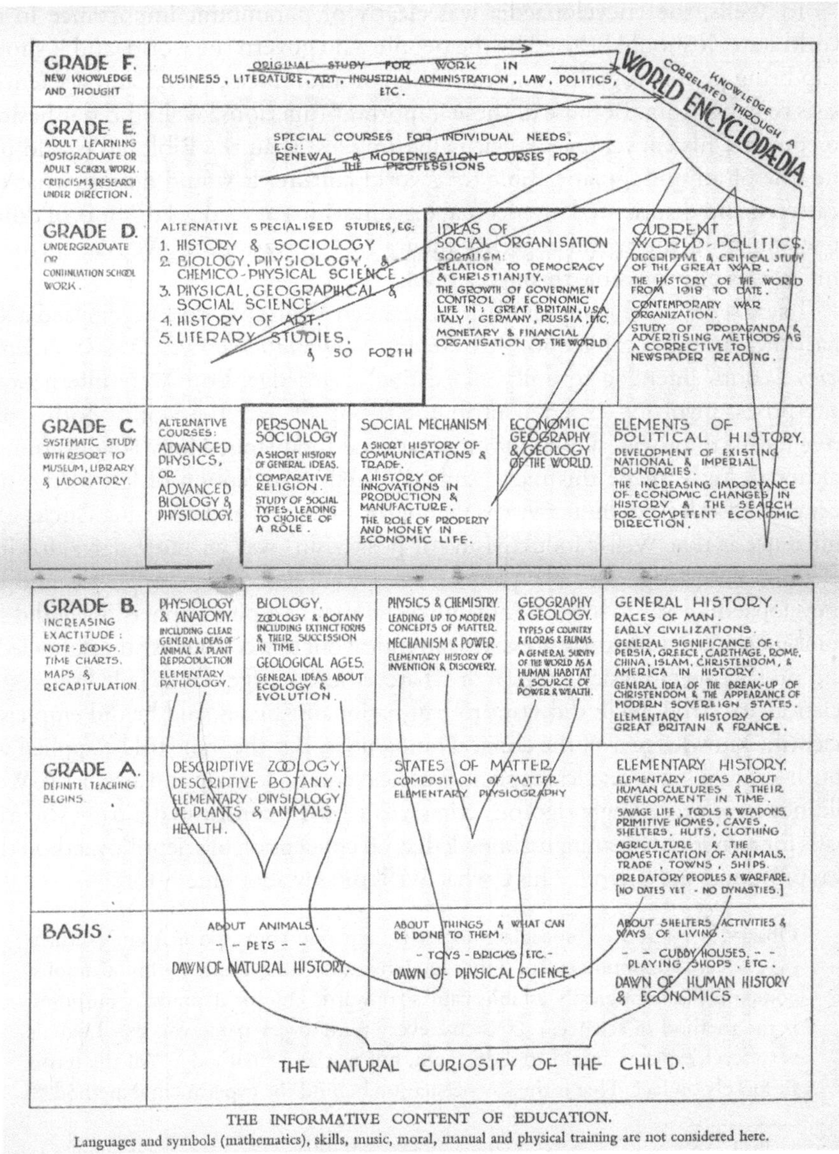

Figure 6.2 Part of the curriculum chart of Wells's new education plan, illuminated by the World Encyclopaedia sun from the top corner.

Source: H.G. Wells, *World Brain* (London, 1938).

feeds, moves, starves, and kills populations' – although the latter activities would cease to be options of choice to the increasingly well-informed world governors.[43]

---

43 Ibid., 23; our emphasis.

To Wells, the encyclopaedia was clearly of paramount importance in the world state. It should be used by the people who govern the world, and it should help bring people together and establish common ideas, values and a scientific basis to work from. Because of these important functions, Wells did not hesitate to compare his encyclopaedia with anything less than the Bible: it 'would play the rôle of an undogmatic Bible to a world culture. It would do just what our scattered and disoriented intellectual organisations of to-day fall short of doing. It would hold the world together mentally'.[44] Salvation, in Wells's scheme of things, was only possible through a book.

There seem to be many similarities between Wells's World Encyclopaedia and that other famous compilation, Otto Neurath's *International Encyclopedia of Unified Science*. Both intended to unify all existing knowledge. Both were international enterprises, involving expert contributors from all over the world. Both works were meant to provide a basis for more rational and planned forms of government (although Wells made this more explicit). Wells's conception of knowledge also seems to be as least as inductivist as that of the members of the Vienna Circle. One difference is that Wells's inductivism (or positivism) was far more naive: for him scientific knowledge was 'like pieces from a complicated jig-saw puzzle' and the encyclopaedia a way to solve it. And unlike the Viennese philosophers, Wells did not combine the compilation process with an endeavour to weed out existing knowledge and rid it of metaphysics, religion or whatever knowledge claims failed the test of scientificity. True, Wells did want to ban nationalist biases, and he did emphasise scientific knowledge over the classical humanities. But the standard he applied was much less developed and less explicit. Unlike Neurath, Carnap and associates, Wells did not have exceedingly rigorous criteria as to what constituted a truly scientific basis for arriving at meaningful knowledge, no conception of scientific method that was profound and sharp.[45] This is what Wells did say about method:

> Observe, try, record, speculate logically, try out your speculation, confirm or correct, communicate to other investigators, hear their communications, compare, discuss logically, establish and so onward. This for all practical purposes, is the method of science.... Distrust every term, every name you use. Logic is very serviceable as an aid to judgement, but not as a final judge. All the terms fit loosely on fact. That is the key persuasion behind the experimental method.[46]

---

44 Ibid., 14.

45 To be sure, not all of the Vienna Circle members always believed in a well-definable scientific method either. Neurath, for example, started to argue against method in the 1930s. Nevertheless, his 'anti-method' views were still more philosophically profound than Wells's views on the subject. See Nancy Cartwright and Jordi Cat, 'Neurath against Method', in R.N. Giere and A.W. Richardson (eds), *Origins of Logical Empiricism* (Minneapolis, 1996), 80–90; and Nancy Cartwright, Jordi Cat, Thomas Uebel and Lola Fleck, *Otto Neurath: Philosophy Between Science and Politics* (Cambridge, 1996), 188–208.

46 Wells, *Outline of Man's Work and Wealth* (n. 31), 67.

Obviously, this is barely an analytical statement of scientific method. It rather looks like a representation of the working practices of scientists as Wells knew them: the ways they worked (or said they worked), and especially the ways they published and talked to each other. Wells was very familiar with many scientific workers, through direct contact, through attending meetings of scientific associations and through his closeness to the editorial work at *Nature*, led by his friend Richard Gregory. It seems that these practices, rather than any profound philosophical reflection, shaped Wells's conception of scientific method.

Wells's world

Something very similar can be said about the 'scientific process' that Wells said should guide world state government. For it still is not entirely clear how exactly he envisaged this process and in what precise ways his world administration would operate. If we are looking for expositions of political procedures or methods of decision making that are somehow scientific, we do not find much in Wells's writings. No planning institutions or directing organisations are ever concretely identified. Instead, what Wells did have to say about the world governing practice was done in brief statements like the one already mentioned: 'World government, like scientific process, will be conducted by statement, criticism, and publication'.[47] As with his statement on scientific method, this barely sounds like political theory. Indeed, it does not sound much like politics at all. What it does come close to is a description of the routines of scientific publishing, the peer review processes that any scientist's contributions had to go through before being accepted and appearing in print. As said before, Wells knew these procedures intimately, if only because overseeing them was the bread and butter of his good friend Gregory. Apparently, Wells believed that these same operations, a similar kind of peer review process, would be the best way to govern the world.

Another association might clarify this. For Wells's process of 'statement, criticism, and publication' not only resembles Gregory's daily activities, it also comes close to a description of the sort of work that Wells himself had been doing, when putting together his outlines of history, biology and so on. These had been no small feats. The writing of each outline had taken one to several years, each had been a multi-author undertaking, and every part was repeatedly checked with appropriate experts.[48] Wells sought and received many comments,

47 Wells, *Open Conspiracy* (n. 27), 31.

48 Wells worked on the *Outline of History* with his wife Jane and a team of six experts; the outline of biology was written with his son George Philip and Julian Huxley; and the social-economic outline was composed in collaboration with Hugh P. Vowles and Edmund Cressy, and later Amber Blanco-White, Odette Keun, A.M. Carr-Saunders and Graham Wallas. Later in the process, several commentators (sometimes dozens of them) were consulted, and Wells would also incorporate comments that came in from readers after publication of the first edition.

which led to various cycles of editing and revision, continuing even after the first version had been published.[49] Hence, the way Wells described the scientific process that would also guide world government ('statement, criticism, and publication') looked very much like the kind of activity that he himself had been involved in for years.

Now this observation explains why an explicit exposition of the scientific process of world government seems to be missing from Wells's work. For this is what that process *was*: making outlines, assembling scientific data, putting knowledge together into large syntheses, just like Wells had been doing in his *Outline of History* and other works and just like he hoped many would do in the making of a World Encyclopaedia. That kind of labour *was* what world government came down to – *it* is what he meant by 'scientific process'. Whenever Wells was asked, 'What do we need to do? How can we head toward global peace? How should the world be governed?', his answer, in one way or the other, would invariably be: start gathering knowledge, start making a World Encyclopaedia. Wells did not envisage the encyclopaedic project as laying the groundwork for world government, but as world government itself. The encyclopaedic organisation was to be something very close to the open conspiracy or the society of samurai that he had pictured before.

Besides the process of government, Wells's conception of the world state looked very much like his own world in other respects as well. His choice, for example, of the language in which its new bible, the World Encyclopaedia, should be written, seems far from coincidental. 'It is a question I have not [completely] worked out', Wells stated, but he also made clear that his preference for the encyclopaedia's most central parts was definitely English, 'because it has a wider range than German, a greater abundance and greater subtlety of expression than French and more precision than Russian'.[50] And with English as its lingua franca, Wells's world state was not altogether unlike the British Commonwealth of his own time. He rejected the notion of empire, with a single mother country subjugating other parts of the globe. But the more egalitarian confederation of Anglophone regions coordinated from one centre that was the Commonwealth did serve him as a model. The 'Encyclopaedic organisation', for example, should be set up that way, 'like a nervous network, a system of mental control about the globe, knitting all the intellectual workers of the world through a common interest and a common medium of expression into a more and more conscious co-operating unity'.[51] In such a network, there was no room

49 Smith, *H.G. Wells* (n. 11), 245–67.

50 Wells, 'World Encyclopaedia' (n. 32), 22. Wells's choice of options seems to reflect his notion of what were scientifically advanced countries. Hence languages that covered larger populations, like Spanish and Mandarin, were no candidates for the World Encyclopaedia.

51 Ibid., 23.

for any kind of regional or local autonomy, Wells pointed out, and the obvious control centre or capital would have to be London.[52]

For operational models of the world state, Wells looked to the British past as well as the British present. One important source of inspiration, mentioned already in *A Modern Utopia* (1905), was the way Francis Bacon had laid out his ideal society in *New Atlantis*. He approved especially of the central role played by Solomon's House, with its fellows in pursuit of knowledge to be used for the benefit and relief of man's estate. In Bacon, Wells recognised his own stress on the practical functions of knowledge and on the notion of a group of 'scientific workers' being the avant-garde leading society.[53] It was no accident, Wells believed, that Bacon had housed this community of world ruling knowledge gatherers in what was in essence an English country home. For it was in those places, he indicated, that genteel culture had grown, preparing the way for the world state:

> Now it is one of my firmest convictions that modern civilisation was begotten and nursed in the households of the prosperous, relatively independent people, the minor nobility, the gentry, and the larger bourgeoisie, which became visibly important in the landscape of the sixteenth century, introducing a new architectural element in the towns, and spreading as country houses and chateaux and villas over the continually more orderly countryside.... Their large rooms, their libraries, their collections of pictures and 'curios' retained into the nineteenth century an atmosphere of unhurried liberal enquiry, of established aesthetic and intellectual standards. Out of such houses came the Royal Society, the *Century of Inventions*, the first museums and laboratories and picture galleries, gentle manners, good writing, and nearly all that is worth while in our civilisation to-day.[54]

---

52 Wells, *Open Conspiracy* (n. 27), 90. For the British Commonwealth as a model for the League of Nations, see Mark Mazower, *No Enchanted Palace: The End of Empire and the Ideological Origins of the United Nations* (Princeton, 2009), ch. 1.

53 Wells, *Modern Utopia* (n. 2), 60, and *Work, Wealth and Happiness of Mankind* (n. 36), 80ff. In the latter, Wells singled out Bacon's notion of a 'research collective' as marking the birth of modern science. Wells's appeals are part of a general tendency in this period to claim Bacon as the herald of the social significance of science. In these attempts, Bacon's renaissance epistemology (which pertained to natural philosophy, law, politics and morals) was reduced to a modern philosophy of science and Bacon was generally recreated in the authors' image. J.D. Bernal, for instance, turned Bacon into a proto-Marxist, writing that 'The ideal of Bacon – the use of science for the welfare of human beings – was indeed a guiding principle of the constructive side of Marxism' – see J.D. Bernal, *The Social Function of Science* (London, 1939), 32. And the (equally Marxist) science writer J.G. Crowther asserted: 'It is evident that the core of Bacon's work was not science, but the social relations of science, and that he was virtually the first, and a very great writer on the subject' – see J.G. Crowther, *The Social Relations of Science* (New York, 1941), 351.

54 H.G. Wells, *Experiment in Autobiography* (New York, 1934), 105. On Wells's idea of genteel culture as the route to the world state, see Kumar, *Utopia and Anti-Utopia* (n. 14), 213.

A PEACEFUL GARDEN IN ENGLAND
Given wisdom, all mankind might live in such gardens

Figure 6.3 Wells's global vision.
Source: H.G. Wells, *A Short History of the World* (London, 1922).

In Wells's future society, this gentle-mannered English class has left their country homes to dominate the globe. Nor would there be any reason to stay on their turf, for transportation technology would take one anywhere in very little time, bridging the greater distances in trains, 'as comfortable as a good club'.[55] Global tourism was to become an important activity in Wells's utopia,

55 Wells, *Modern Utopia* (n. 2), 46.

especially since '[t]he greater part of the world will be as secure and cheaply and easily accessible to everyone as is Zermatt or Lucerne to a Western European of the middle-class at the present time'.[56] There will be convenient inns everywhere and there will be cycle tracks 'amidst woods and crops and pastures'.[57] Even the landscape seems to have been imported from the English countryside:

> There will be pleasant ways over the scented needles of the mountain pinewoods, primrose-strewn tracks amidst the budding thickets of the lower country, paths running beside rushing streams, paths across the wide spaces of the corn land, and, above all, paths through the flowery garden spaces amidst which the houses in the towns will stand. And everywhere about the world, on road and on path, by sea and land, the happy holiday Utopians will go.[58]

As the picture of the garden and its caption from one of Wells's books further testify (Figure 6.3), it was these English upper middle-class luxuries, which Wells himself had acquired from humble beginnings, that he now wished for the whole world to enjoy. If it could only follow his personal development of upward mobility through education, all of humanity might reach a similar position. The future society was never far away for Wells. From its system of government to its choice of language, from its conception of science right down to its landscape, it reflected the world as he knew it. For all its universalism and global-mindedness, Wells's world state never strayed far from his own particular circumstances.

Acknowledgement

We thank Charlotte Thomsen, for presenting the first version of this paper at the workshop in Maastricht, and Harmke Kamminga for constructive suggestions and substantive advice.

---

56 Ibid., 44.

57 Ibid., 44 and 47.

58 Ibid., 47.

Chapter 7

Unifying science against fascism

Neuropsychiatry and medical education in the Spanish Civil War

Fernando Salmón and Rafael Huertas

In this chapter we explore the use of science as a key weapon in the rhetoric that sustained the defence of the Second Republic during the Spanish Civil War (1936–1939). In its own propaganda, the Spanish Republic was projected as the guardian of the highest values of modern civilisation, and science was used as a unified category that incorporated and reinforced republican values and goals.

Already from its start in 1931, the newborn republic included science in its identification with secularism. Scientists saw in the new regime the opportunity to widen the significance of their enterprise as an ordering principle for the whole of society. Within this context the differences between particular disciplines became irrelevant and the discussion of their boundaries was considered futile. Science was one and, as such, represented not only technical progress and modernity but also, importantly, an attitude and an instrument in the forging of new social relationships.

Within this general construction of science as republicanism, we focus on the prominent role played by the *Asociación Española de Neuropsiquiatría* (Spanish Association of Neuropsychiatry, AEN) in developing specific projects for putting science at the heart of a republican social order. While the Civil War put a halt to these aspirations, some individual initiatives were followed up in cities that remained loyal to the republic. We analyse two projects established in northern Spain by a leading member of the AEN, Wenceslao López Albo (1889–1944), which were strategically developed to serve the republican cause by identifying science and republic: a Free School of Medicine and a farm-asylum as part of an ambitious plan for the study and care of the mentally ill. Inaugurated in Santander and Solares in November 1936 and January 1937, respectively, they were closed down after the entry of Franco's troops in August 1937. Although they were short-lived, these institutions had a strong impact on republican morale at the time.[1]

---

1 See e.g. H. Tellez Plasencia, 'Problemas Culturales de Santander. La Universidad Internacional', *El Cantábrico* (hereafter *EC*), 26 January 1937. Some of the themes developed in this chapter were earlier analysed more briefly in, F. Salmón, 'Proyectos de la Razón:

Yet historians have ignored the projects as if they had never existed.[2] Our work aims to restore them to memory and give them historical significance, above all as attempts to unite science with a particular view of society.

Science and republic

The union between science and social reform, science and republic and also between science and revolution was a constant feature in Spanish discourse about modernisation from the 1870s to the 1930s. In pursuit of modernisation, reformers strove for the transformation of Spain from a reactionary monarchy allied with the Catholic Church into a progressive secular republic. Simultaneously, they saw science as the driving force of intellectual and social progress, thus linking scientific and republican values.[3]

Although the new constitution of 1876 had limited the powers of the Crown and created a two-chamber parliamentary government, parliament never gained enough strength to steer post-imperial Spain towards the status of a modern European nation state, nor to stimulate the expansion of scientific research to the extent deemed necessary. In the early twentieth century, widespread discontent with the lack of social and economic progress led to deep political divisions and a proliferation of political movements.

Eventually disillusionment with the weak parliament gave General Primo de Rivera sufficient support for a military coup in 1923. The political stability imposed under his dictatorship made room for some material gains and programmes of social reform, but republican ideals remained strong. Despite press censorship and repression of demonstrations, the intellectual elite and academics were overwhelmingly and openly against the new regime and the monarchy that supported it. In 1930, having lost the confidence of his military fellows and the king, Primo was forced to resign and died in exile a few months later. He was replaced by General Berenguer, under whose rule the 1931 municipal elections took place that were won by the anti-monarchist parties. Their victory forced the king to abdicate, and the Second Republic was declared. The republican

Práctica y Retórica Científica en la Colonia Agrícola para Alienados y en la Facultad de Medicina de la Casa de Salud Valdecilla de Santander (1936–1937)', *Revista de Historia de la Psicología* 26 (2005), 272–92.

2 It is striking that not a single reference to the Free School of Medicine or the farm-asylum appeared in the commemorative volume devoted to the Valdecilla Hospital in 1999, not even in the chapters on López Albo and on the hospital during the Second Republic and the Civil War. See J.M. Izquierdo Rojo (ed.), *70 Años de Valdecilla. Salud y Prestigio para Cantabria. Algo Más que un Hospital* (Santander, 1999), esp. 49–59 and 63–77.

3 M. Suárez Cortina, *El Gorro Frigio. Liberalismo, Democracia y Republicanismo en la Restauración* (Madrid, 2001).

government was divided internally, however, and Spanish politics became increasingly polarised. The left-wing Popular Front alliance narrowly won the general election of February 1936, but within six months military insurrections against the government, instigated among others by General Francisco Franco, set off the infamous Civil War. Despite unprecedented international support for the republican cause, Franco's rebel army, with massive material assistance from Fascist Italy and Nazi Germany, prevailed.[4]

It was against this background that Spanish scientists and intellectuals in general campaigned for the identification of the values of the cultural-scientific enterprise with republican values.[5] As we will discuss, of all professional bodies, the *Asociación Española de Neuropsiquiatría* (Spanish Association of Neuropsychiatry, AEN) was especially active in promoting the science/republic relationship. Constituted formally in 1924, the AEN not only enthusiastically hailed the establishment of the Second Republic seven years later, but few bodies were as successful in transforming this enthusiasm into concrete scientific and health policies.[6]

Republican politicians recognised the authority of the AEN and its representatives. Memories were still alive of the eulogy uttered by Pérez Galdós upon the death of José María Esquerdo (1842–1912), a pillar of modern Spanish psychiatry: 'Apostle and leader of the two religions: Science and Republic'.[7] However, the AEN did not primarily owe its authority among republican politicians to a long involvement of well-known psychiatrists in the promotion of the republican cause, but above all to the image of the scientist as patriot of a new Spain. In this regard the status of Santiago Ramón y Cajal (1852–1934) exercised a unique influence.[8] The first – and for many decades the only – Spanish Nobel laureate in the sciences (for physiology and medicine, 1906), Cajal became a symbol of Spanish modernisation, both scientific and social,

---

4 For a general view, see R. Carr, *The Spanish Tragedy: The Civil War in Perspective* (London, 1977).

5 F.A. González Redondo, 'Los Científicos Españoles entre la República *en Paz* y la República *en Guerra*. Compromisos y Desafecciones', in A. Gómez Rodríguez and A.F. Canales Serrano (eds), *Estudios Políticos de la Ciencia. Políticas y Desarrollo Científico en el Siglo XX* (México, 2013), 53–80.

6 R. Huertas, *Organizar y Persuadir. Estrategias Profesionales y Retóricas de Legitimación de la Medicina Mental Española (1875–1936)* (Madrid, 2002), 107–24 and 169–212.

7 C. Llorca, *José María Esquerdo: el Gran Desconocido* (Villajoyosa, 1984), 131–2.

8 For a bibliography of works by and on Cajal, see J.M. López Piñero, M.L. Terrada and A. Rodríguez Quiroga, *Bibliografía Cajaliana. Ediciones de los Escritos de Santiago Ramón y Cajal y Estudios sobre su Vida y su Obra* (Valencia, 2000). See also A. Albarracín Teulón, *Santiago Ramón y Cajal o la Pasión por España* (Barcelona, 1982); and J.M. López Piñero, *Cajal* (Madrid, 2000).

to which the AEN appealed. Several outstanding members of the AEN were trained in neurohistological research under Cajal's direct supervision or sponsorship. The new political circumstances of 1931 allowed the members of the AEN to use this connection far outside the walls of the laboratory. They did not need to introduce themselves to republican politicians, explaining first why scientists had a voice to be listened to in the design of a new state. Cajal had already done it for them.

Cajal's influence was important in two ways. First, he legitimised specific political proposals of a Krausist nature,[9] which were favoured and adopted by the Second Republic, in his role as President of the *Junta para Amplicación de Estudios e Investigaciones Científicas* (Board for Expansion of Scientific Study and Research, JAE). The activities of the JAE (founded in 1907), with its programme of grants for study abroad, had a clear impact on medicine during the interwar years, and especially on neuropsychiatry, which was not part of the syllabus in Spanish Faculties of Medicine.[10] Secondly, throughout the first third of the twentieth century, the figure of Cajal was taken as the incarnation of the ideal scientist. Often portrayed with a microscope, and situated carefully outside specific political affiliations and party interests (e.g. his rejection of remuneration as president of the JAE was well-known), he was bestowed with a moral authority that extended to questions far beyond the limits of the laboratory (Figure 7.1).[11] And conscious of the importance of his own persona, Cajal meticulously supplied the materials for its construction, both in his scientific work and in his essays and autobiographical accounts.[12]

---

9 The philosophy of Karl K.F. Krause (1781–1832) had a strong impact in Spain in the late nineteenth century through the *Institución Libre de Enseñanza* (Institution of Free Learning) as a pedagogical initiative that aimed at ambitious cultural reform of the country based on secular values. See A. Jiménez García, *El Krausismo y la Institución Libre de Enseñanza* (Madrid, 1986).

10 On the activities and ideology of the JAE, see *Arbor. Ciencia, Pensamiento y Cultura* 126 (issue 493) (1987): a special issue on 'Ochenta Años de Ciencia en España. Los Orígenes Culturales de la Junta para Ampliación de Estudios'. See also J.M. Sánchez Ron (ed.), *La Junta para Ampliación de Estudios e Investigaciones Científicas 80 años después* (Madrid, 1988), 2 vols. See also, *Asclepio* 59 (2) (2007): a special issue on the occasion of the centenary of the JAE. Cajal himself provided an account of the activities of the JAE under his presidency in S. Ramón y Cajal, 'La Investigación Científica en España durante los Últimos Veinticinco Años', in *España Bajo el Reinado de Alfonso XIII. Veinticinco Años de Paz, Cultura y Progreso (1902–1927)* (Barcelona, 1927), 55–60.

11 Besides being non-remunerative Cajal had also demanded that the presidency of the JAE be strictly independent of any changes in the political colour of the government. E. Lewy Rodriguez, *Santiago Ramón y Cajal* (Madrid, 1987), 98.

12 For example, S. Ramón y Cajal, *Recuerdos de mi Vida* (Madrid, 1923).

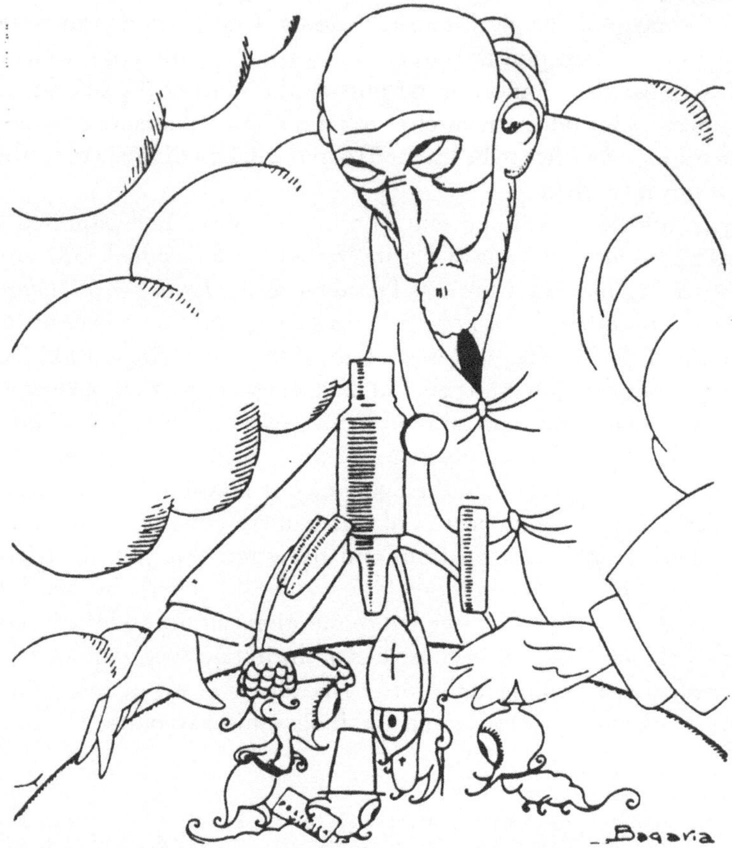

Figure 7.1 The political cartoonist Bagaría portrays Cajal, the scientist, inspecting the ruling powers of Spanish society.

Source: La Nación, Buenos Aires, 29 June 1924. Reproduced with permission from the Fundación Mapfre Art Collection.

There was, however, little public awareness of Cajal's histological investigations of the nervous system. They neither promised quasi-miraculous cures, nor produced technologies of the economic significance that, as in the case of Pasteur, might explain popular glorification of the scientist. What did Cajal offer? His own self. The scientist Cajal was shaping up as an apostle of the new secular religion.[13] The promotion of science as a way of life and of the scientist as a socially responsible citizen with high moral values was

---

13 López Piñero has criticised this image as being the result of a later historical falsification. López Piñero, *Cajal* (n. 8), 165–211. However, Cajal and his contemporaries made ample use of themes drawn from religious asceticism to portray the life of the scientist.

more powerful than understanding the scope of his research or its place within specific disciplinary boundaries. In this sense, Cajal's precise area of knowledge and research was irrelevant. Science as a secular religion did not need to be explained, but just worshipped. As Ortega y Gasset, influential philosopher and contemporary of Cajal, denounced bitterly, the interest shown by Spanish society in scientific research was without substance and was fulfilled simply by having at hand a Nobel laureate who could be exhibited 'like the body of Saint Isidro, as a magical fetish. . . . Because scarcely anyone has the slightest idea of the admirable conquests of such an illustrious scholar'.[14]

This image of the man of science as an exceptional genius or secular saint was instrumental in ensuring that Cajal's voice was heard with authority in fora not strictly scientific. But it had to be tempered for the project of regeneration of the nation that Cajal championed. To this end, conscious of the dangers of seeming exceptional, Cajal insistently described himself as a person of humble origins, without outstanding qualities and endowed with average intelligence. The example he thus set was not too far removed from the experience of the majority, so that young people could be encouraged to emulate him. Such emulation would bring about, not geniuses, but a critical mass of scientific workers, the only means by which a modernised Spain could again have a role on the world stage.

The moral and material regeneration of the country through science was seen as an attitude and patriotic duty rather than a set of particular practices, and this was a constant theme in Cajal's writings aimed at awakening the consciousness of Spanish youth.[15] In fact, some contemporaries felt his promotion of modernisation through science to be more important than the histological research that won him the Nobel Prize. Gregorio Marañón, socially and scientifically the most influential clinician of the interwar generation, also expressed this view after Cajal's death in 1934. In the memorial session held by the Royal Academy of Medicine, Marañón called Cajal 'the distinguished master of us all – scientists and professionals, doctors and non-medics'. He went on to state that for all Spanish citizens 'being Spanish is enough to feel that we are disciples of his teaching about the means of recovering the dignity of the fatherland, a pursuit in which he was perhaps

---

[14] C. Castilla del Pino, 'Cajal: las Razones de un Mito', in E. Lewy Rodríguez (ed.), *Así era Cajal* (Madrid, 1977), 14–15. San Isidro was the patron saint to whom Madrid dedicated its most popular open-air celebration (mass, dance and bull-fighting included). Here it is used by an elitist Ortega to highlight the cultural poverty of contemporary Spanish society.

[15] See, for example, the postscript that Cajal added in 1899 to his *Los Tónicos de la Voluntad* (Madrid, 1982), 177–86. On science and the regeneration effort, see E. García Camarero, *La Polémica de la Ciencia Española* (Madrid, 1979).

even more effective and distinguished than in his brilliant anatomical and biological work'.[16]

'Fatherland and science' was easily converted into 'republic and science' for a later generation of scientists who strongly supported the proclamation of the Second Republic. This applied especially to neuropsychiatrists, who could collectively claim Cajal's unique status for themselves so as to be seen as his natural heirs. Most of them had training that combined detailed neurohistological research with clinical neurology and psychiatry, and with this background they sought to embody both the authority of the scientist and the social influence of the medical professional. On this stance the neuropsychiatrists portrayed their potential fields of intervention far beyond those of older, more established medical specialisms. To this end they established themselves as a professional group in the AEN in 1924 and in the *Liga de Higiene Mental* (League of Mental Hygiene) in 1926. These bodies worked on proposals for psychiatric reform that the new republican administration immediately adopted as its own.[17] In July 1931, barely three months after the declaration of the Second Republic, it decreed a mental care reform that radically changed the legislation that had been in force since 1885. This new decree opened the door to psychiatrists acting and advising within a wider scheme of social planning, not just on mental health policy in a narrow sense.[18]

From then onwards, an entirely new administrative structure was to be created for all psychiatric legal proceedings throughout the nation, under the control of the *Dirección General de Sanidad* (General Health Administration). This structure would rest on two pillars: the *Consejo Superior Psiquiátrico* (Higher Psychiatric Council), with advisory functions of planning and inspection, and the *Sección Central de Psiquiatría e Higiene Mental* (Central Department of Psychiatry and Mental Hygiene) which, together with *Secciones*

---

16 G. Marañón, 'Recuerdo a Cajal', speech to the Royal Academy of Medicine, 29 October 1934, in G. Marañón, *Obras Completas*, vol. 2 (Madrid, 1971), 315. Cajal himself encouraged this notion. For example, in his reply to the homage that medical students offered him in 1931, Cajal said: 'If the Ego was not something loathsome and if what I am going to say would not be understood as a sign of childish vanity, I would declare that in me it is only fair to praise the patriot'. Anon., 'El Homenaje a Cajal en la Facultad de Medicina de Madrid', *Gaceta Médica Española*, 5 (9) (1931), 264-p. From here onwards all references with '-p' in the page numbering refer to the 'Sección Paramédica' of the *Gaceta Médica Española*.

17 J. Lázaro, 'La Fundación de la Liga Española de Higiene Mental', in J. Castellanos, L. Jiménez Lucena and M.J. Ruiz Somavilla (eds), *La Medicina en el Siglo XX. Estudios Históricos sobre Medicina, Sociedad y Estado* (Málaga, 1998), 213–20; idem, 'Historia de la Asociación Española de Neuropsiquiatría (1924–1999)', *Revista de la Asociación Española de Neuropsiquiatría* 20 (2000), 397–515.

18 Decree of 3 July 1931, official publication in *Gaceta*, 7 July 1931.

Provinciales (Provincial Sections), would have direct administrative and executive functions.[19]

It is no coincidence that Cajal was offered the post of Honorary President of the Higher Psychiatric Council. Without commenting openly on the new regime, Cajal permitted the use of his name in initiatives of professionals with whom he had close connections. Already in 1920 he had agreed to preside over the editorial committee of the *Archivos de Neurobiología* (Neurobiology Archives), a journal that would later become the official organ of the AEN.[20] The AEN adopted a strongly somatic orientation to psychiatry, following the German model, and was fully cognisant of the newest neurohistological research. Many neuropsychiatrists had worked in Cajal's Institute of Biological Research or had benefited from the JAE's grants to study abroad. By the 1930s recognition of the profession and authority of neuropsychiatrists was a reality. The AEN had an influential voice, whether it presented laboratory findings or proposed plans of social engineering.[21]

Science and modernity as secular values of a new republican Spain found an effective place in the social intervention proposals of the AEN. It is clear that the decree of 1931 was not a hurried and improvised proposal of the new government. Its text corresponded to earlier work presented by prominent members of the AEN to the Third Assembly of the League of Mental Hygiene, held in Seville in 1929, still under the dictatorship of Primo de Rivera. When this plan was approved, the AEN turned it into a legislative and political initiative for which it campaigned forcefully.[22]

The republican administration not only listened to the neuropsychiatrists, but also expected them to set up the organs of government that had to implement the reforms.[23] It is not surprising, therefore, that the republic in turn received the unconditional support of most of the membership of the AEN. Wenceslao López Albo, president of the AEN at the time of the first meeting held under a republican government (Granada, October 1932), recognised in his inaugural

---

19 Huertas, *Organizar y Persuadir* (n. 6), 180–84.

20 D. Gracia, 'Medio Siglo de Psiquiatría Española: 1885–1936', *Cuadernos de Historia de la Medicina Española* 10 (1971), 305–39.

21 Gonzalo Rodriguez Lafora, one of the founding members of the AEN, is a good example. See L. Valenciano Gayá, *El doctor Lafora y su Época* (Madrid, 1977).

22 Huertas, *Organizar y Persuadir* (n. 6), 180–84. See also R. Huertas and R. Campos, 'La Medicina Mental en la Segunda República Española. Retóricas Legitimadoras e Iniciativas Políticas', in A. Gómez Rodríguez and A.F. Canales Serrano (eds), *Estudios Políticos de la Ciencia. Políticas y Desarrollo Científico en el Siglo XX* (México, 2013), 81–103.

23 Lázaro, 'Historia de la Asociación' (n. 17), 429. See also R. Álvarez Peláez, 'La Psiquiatría Española en la Segunda República', in *Un siglo de psiquiatría en España* (Madrid, 1995), pp. 89–111.

speech that some of the members had temporarily abandoned their scientific activities in order to help set up the new regime.[24] In 1935, the work of the Higher Psychiatric Council resulted, among other things, in a *Proyecto de Ley de Organización de la Asistencia Psiquiátrica Nacional* (Draft Law for the Organisation of National Psychiatric Care), presented in parliament by the Minister of Work, Health and Social Security. The draft envisaged that the state would have full responsibility for psychiatric care and assumed that the mental patient should be considered, not merely from a legal viewpoint of rights and enforcements, but from a scientific perspective involving clinical research and monitoring of treatment. It also stressed the need to set up the psychiatric dispensaries already proposed in the 1931 decree, as a means of preventing mental illness and for the organisation of social welfare and wider social intervention.[25] The military insurrection of 1936 prevented the implementation of the law, although parts of it were put into practice in the republican area during the Civil War.

Science and republican war propaganda

In January 1937 López Albo inaugurated a farm-asylum for the mentally ill in Solares, Cantabria. It was part of a general plan of care for mental patients that brought together the three proposals of the draft law: care for outpatients in mental hygiene dispensaries, active treatment in the psychiatric hospital and care for chronic patients in psychiatric farm-asylums.[26] Barely two months earlier, in November 1936, López Albo had made the inaugural speech in the *Casa de Salud Valdecilla* (Valdecilla Hospital) in Santander, at the opening of a Free School of Medicine.[27] (The two projects will be discussed in detail later.)

The historical significance of these projects should not only be understood in terms of their practical value, but even more so in the way they were used for propaganda purposes in the Civil War. Their practical significance was obvious, albeit constrained by contemporary conditions. The farm-asylum made possible the discharge of chronic patients from the hospital's psychiatric unit, which could therefore focus on acute cases. The Free School of Medicine gave medical

---

24 W. López Albo, 'Discurso del Presidente. Sexta Reunión anual de la Asociación Española de Neuropsiquiatras (Granada, 2, 3, 4 y 5 de Octubre de 1932)', *Archivos de Neurobiología* 12 (1932), 947.

25 Huertas, *Organizar y Persuadir* (n. 6), 208–10.

26 *EC*, 2 February 1937.

27 *La Voz de Cantabria* (hereafter *LVC*), 24 November 1936; *El Diario Montañés* (hereafter *EDM*), 24 November 1936; *EC*, 24 November 1936.

students who were in Santander at the start of the Civil War the opportunity of continuing their training.

Both projects also served important propaganda purposes. The Free School of Medicine and the regional plan for care of the mentally ill, of which the farm-asylum was only a part, embodied in a very direct manner republican interest in science and culture. The recognition that science and republic shared values of secularism, modernity and progress provided powerful elements of identification that were extremely useful in a situation of war. In fact, the wartime publicity given by the press to the two projects generated a sense of local cultural euphoria, leading to other proposals that were harder to realise, such as the creation of a Provincial Museum, a Faculty of Fine Arts and the conversion of the International Summer University into the University of Cantabria.[28]

The propaganda discourse that assimilated science and republic rested on a complex connection of words and images in which the Second Republic appeared as the zealous guardian of the highest values of western civilisation.[29] Commitment to access to higher education, and fighting illiteracy as its necessary precondition, was a constant theme in republican propaganda and in actual programmes during the war (Figures 7.2 and 7.3). The press emphasised the association of republic to civilisation by outlining, in stark contrast, the barbarity and contempt for culture that drove Franco's rebel army. This was evident, for example, in the way the news was presented of the fascist bombardment of the university campus of Madrid and of the *Casa de Veláquez,* a centre for advanced research.[30] The identification of the republic with science and civilisation, and the contrast to the rebel army's cry of '*Muera la Inteligencia!*' ('Death to Intelligence!'), were among the most persistent themes of republican writing produced in the Civil War.

The narrative framework that sustained this discourse also made strong use of history. Both the Free School of Medicine and the farm-asylum were endowed with a history that legitimised their creation and countered any sense of provisionality. Presenting them also as projects for future development served to reinforce the suggestion of stability, strengthening confidence, at a time of civil war, in the continuity and flourishing of the political, social and

28 For example, Téllez Plasencia, 'Problemas Culturales' (n. 1), and idem, 'Más sobre la Universidad Internacional' *EC*, 21 February 1937.

29 Republican propaganda effectively fuelled hatred of the 'Moors' by highlighting the presence of north African soldiers in the rebel army and by presenting this as a threat to genuine Spanish culture. See F.J. Pérez Rojas, J.L. Alcaide, M. Escriche (eds), *Arte y Propaganda. Carteles de la Universitat de València* (Valencia, 2002), 136–7.

30 For example, E. Noreña, 'La Ciudad Universitaria de Santander', *EC*, 27 November 1936.

Figure 7.2 Poster from the Ministry of Public Education (Madrid, 1937), stating: 'In 1937, 5 million pesetas are being dedicated to enable the children of antifascist fighters to attend universities and technical schools'.

Source: Collection of the Fundación Pablo Iglesias. Reproduced with permission.

intellectual project of the republic. Accordingly, the Free School of Medicine was presented as the first phase of the University of Cantabria, which, after the conflict, would incorporate a set of related institutions in Santander – among

Figure 7.3 Poster from the Madrid Provincial Union of Schoolteachers (Madrid, 1937) promoting the fight against illiteracy. It is headed, 'The ignorant worker forges his own chains' and advises 'Workers unions: let there not be a single illiterate among your members'.

Source: Collection of the Fundación Pablo Iglesias. Reproduced with permission.

them the International University, the Menéndez Pelayo Library, the Marine Biological Research Station and the Valdecilla Hospital.[31] As López Albo wrote

[31] The same line of thought informed the proposal presented in 1935 by López Albo for the creation of a School of Medicine. W. López Albo, 'Porvenir de Santander como Ciudad Universitaria', *EC*, 17 August 1935.

in November 1936, the envisaged university would be linked to the autonomous region that would be part of the federal regime of 'the new Spain, the true Spain, that will emerge with the triumph of universal democracy over international fascism. . . . [T]hat which is born today as the School of Medicine will then just be a part of the University of Cantabria'.[32]

In a similar manner, the farm-asylum was presented as one of the services offered within the plan of regional psychiatric care that López Albo had proposed ten years earlier and which had already led to several extensions of the Valdecilla Hospital.[33] The project of medical teaching and research and psychiatric care was hence associated with what was already a quite exceptional provincial charitable hospital – in fact one of the most powerful medical-scientific and social icons developed during the final years of the dictatorship of Primo de Rivera.

The Valdecilla Hospital and Wenceslao López Albo

The Valdecilla Hospital project had begun in 1927, against a background of attempts by the Cantabria *Diputación* (provincial assembly) and by private associations to replace the old charitable hospital of San Rafael with one that would offer care to the poor and the industrial proletariat of the province. For various reasons, these attempts failed until the president of the assembly enlisted Ramón Pelayo, first marquis of Valdecilla (1850–1932), in the project in 1926. This philanthropist was a wealthy businessman who had made his fortune in Cuba. Since his return to Spain in 1917, he had dedicated a large part of his capital to social and cultural projects – which brought him the friendship of the royal family, a marquisate, the respect of the dictator and popular admiration. His projects were ideological expressions of his personal interpretation of the main planks of the regenerationist movement aimed at modernising Spain: education and health. To this ideology he added the values of North American philanthropic culture, with which he had come into contact while trading in Cuba.

The marquis provided the money for the new hospital and, in effect, took over the project, shaping it as an institution for medical care, teaching and research. It meant that a private initiative adopted the functions of a general hospital in a provincial environment, thus creating an autonomous space that could control the public services that otherwise relied on the dictatorship's

32 *LVC*, 24 November 1936.

33 By 1919 a pavilion for the acutely mentally ill and an outpatient service had been realised. W. López Albo, 'Plan Moderno de Asistencia a los Alienados', *Archivos de Neurobiología* 6 (1926), 219–26. See also F. Salmón, L. García Ballester and J. Arrizabalaga, *La Casa de Salud Valdecilla. Origen y Antecedentes. La Introducción del Hospital Contemporáneo en España* (Santander, 1990), 161–77.

administration through the provincial assembly.[34] In April 1928, Ramón Pelayo, advised by Gregorio Marañón, entrusted López Albo with the scientific organisation and management of the Valdecilla Hospital and appointed him head of the psychiatric pavilion.[35] At the time of his appointment, López Albo was director of the Mental Asylum of Zaldívar and head of neuropsychiatric surgery in the Civil Hospital of Bilbao.

Hardly any archival material about López Albo remains. His private papers were lost from the family archives upon his exile during Franco's regime. That same regime carefully expurgated the archives of the Valdecilla Hospital. Furthermore, the heirs of the marquis of Valdecilla do not allow access to his archives. The sources for reconstructing the professional career of López Albo are therefore almost exclusively biographical notes in the medical press accompanying presentations of his research. But the problem of interpreting portrayals of this type is compensated by the advantages of obtaining a professional profile that answers to an ideal type.[36] For López Albo's career corresponded to that of a generation of neuropsychiatrists, characterised by Marañón in 1933 as such: education and specialisation outside official channels, extensive knowledge of languages, postgraduate training in scientific institutions abroad, attention to international scientific developments, promoting foreign works by writing reviews and translations, emphasising experimental and clinical research and having a special awareness of social problems.[37] Ideologically, this generation exhibited an optimistic confidence in science as an agent of modernisation.

The coming together of Ramón Pelayo as sponsor and López Albo as organiser allowed the incorporation of this spirit in the design of the Valdecilla Hospital. In October 1929 the surgeries of the hospital were opened to the public, and in January 1930 its Postgraduate Medical Institute was inaugurated. The Board of Patrons of the Valdecilla Hospital, under the chairmanship of the marquis, respected the independence of López Albo as director of the hospital and gave him full scientific and administrative control.

The Valdecilla Hospital project was the most important professional achievement of López Albo's life. An active member of the medical section of the Basque *Unión General de Trabajadores* (General Workers Union UGT) and close to radical socialism with its anticlericalism, he believed in the power of science as a tool of social transformation and as a liberating force in a society

34 Ibid., 41–96.

35 *Libro de Actas del Patronato de la Casa de Salud Valdecilla* (hereafter *LACSV*), 16 April 1928.

36 For a biography of López Albo, see F. Salmón, 'Wenceslao López Albo y los Orígenes de la Casa de Salud Valdecilla (1928–1930). Filantropía y Medicina en la España Contemporánea' (MD thesis, University of Cantabria, 1986), 380–90.

37 G. Marañón, *Raíz y Decoro de España* (Madrid, 1933).

dominated by the church.[38] In the organisation of the new hospital, he saw opportunities for putting into practice his ideals for a just social order, ruled by scientific principles.[39] Most of the innovations that were incorporated in the hospital routine reflected this ideal and received the powerful sanction of being labelled scientific. From the hospital's administrative organisation to its dietetic practices, 'scientific' and 'modern' were the adjectives that accompanied their descriptions in the daily and professional press. Among the changes that departed from traditional ways of hospital organisation, the one that met stronger opposition was the absence of a religious order as part of the hospital staff. The decision was justified on professional and scientific grounds since most nuns had no formal training as nurses or administrators. It soon became clear that the Board of Patrons barely tolerated this move and was quick to change it at the first opportunity.

The programme that López Albo designed envisaged four functions for the Valdecilla Hospital: health care for all social classes, teaching of medical internees and nurses, clinical and experimental research, and preventive social care. The organisation and structure of the Valdecilla Hospital were inspired by discussions of the American Hospital Association and included the latest scientific models and materials exhibited at meetings of the International Hospital Association. López Albo's proposals should be understood in terms of the attention he paid to this international movement as well as national debates about medical teaching, research and health care. Indeed, the daily and professional press outlined the striking peculiarities of the hospital as unique to Spanish hospital culture.[40]

The realisation of these proposals would not have been possible without the scientific and economic autonomy López Albo enjoyed. In 1930, however, the Board had a change of heart and assumed an interventionist stance that, within a few months, led to a complete reorganisation of the original set-up. The marquis, already infirm and confined to his home, lost control of his project. The reorganisation took shape with the admission of a community of nuns that was placed in charge of all nursing and administrative duties. In practice, this meant that López Albo, as medical and administrative director, lost financial control, with technical and professional decisions becoming subordinate to economic criteria imposed by the Board through the Superior of the Sisters of Charity. The change was meant as a clear political message, demonstrating the power of the local conservative oligarchy and of the bishop, an ex officio

38 Anon., 'Sindicato Médico de Vizcaya', *Gaceta Médica Española* 6 (1931), 29-p.

39 For example, W. López Albo, *Planificación y Organización de un Hospital Moderno* (San Sebastián, 1932).

40 See the special issue that *Gaceta Médica Española* (4, 1930) devoted to the Valdecilla Hospital.

member of the Board. Resignations were not long in coming: Teresa Junquera, Assistant Director of the Nursing School; Matilde Basterra, Head of the Social Services; Miguel Caballero, Managing Director of the Hospital; and finally, in September 1930, López Albo himself.[41]

After giving up the directorship of the Valdecilla Hospital, López Albo returned to his private practice in Bilbao, where he remained until September 1936 when he was summoned to resume the directorship of the Valdecilla Hospital.[42] When the entry of pro-Franco troops into Santander was imminent in August 1937, he was forced to leave the city hurriedly by sea to France. From there he travelled to Barcelona, where he became professor at the *Universidad Autónoma* (Autonomous University) in 1937–38 and occupied various responsible posts in the republican organisation of health services.[43] When Barcelona fell in 1939, he went into exile in Mexico. There he did clinical work, teaching and research, first in Monterrey and from 1942 in Mexico City, where he died on 28 December 1944.[44]

The Free School of Medicine

When López Albo returned to Santander in September 1936 to take charge again of the Valdecilla Hospital and its psychiatric pavilion, he did so at the request of the socialist Bruno Alonso, who, as the regional minister of war, was in charge of organising all public services.[45] Under these circumstances, more radical innovations could be made.[46]

Already in October 1936 press advertisements had offered medical students the opportunity of participating in the scientific life of the institution.[47] Several newspapers characterised the proposal as the de facto creation of a 'Free School of Medicine' that was to become an official Faculty of Medicine at the cessation

---

41 Salmón, García Ballester and Arrizabalaga, *Casa de Salud* (n. 33), 257–76.

42 Anon., 'Cambio de dirección en la Casa de Salud Valdecilla' *Gaceta Médica Española* 5 (1930), 7-p.; *LACSV*, 10 September 1930. *EC*, 26 September 1936.

43 López Albo was head of the neurosurgery service of the military hospital at Barcelona (1937–39), neurology advisor to the *Dirección General de Sanidad* (DGS, Department of Health) of the Army (1937–38), head of health institutions of the DGS (1937–38), and president of the neuropsychiatry and neurosurgery tribunal of the DGS up to 1939. This information is contained in the few extant documents kept by López Albo's family.

44 J.M. Izquierdo Rojo, *Historia de la Neurología Clínica Española (1882–1936)* (Oviedo, 1978), 221.

45 *EC*, 26 September 1936.

46 *EC*, 26 September 1936.

47 *EC*, 22 October 1936.

of the conflict.[48] While the school did not have formal legal standing, the presence of all regional politicians at the opening ceremony on Sunday 22 November clearly signalled official approval of the project.

López Albo was entrusted to deliver the inaugural address, which received ample publicity.[49] There are interesting parallels between this address and his speech at the inauguration of the hospital's Postgraduate Medical Institute in 1930.[50] Both were delivered in uncertain political and social circumstances – the first in the final months of Primo de Rivera's dictatorship, the second in wartime. Both emphasised the value of science and culture as tools of social justice. In the first, López Albo stressed the teaching of professional scientists with a commitment to care for all social classes 'in order to procure the advance of medicine and to do social and scientific work beyond the narrow and self-interested confines of professionalism'.[51] In the second, he emphasised the right to a university education for all capable people regardless of social origin, a commitment taken directly from the text of the Republican Constitution.[52] The tone of López Albo's speech in 1936 deliberately avoided the passion of political rhetoric. Instead, he stressed the continuity of the scientific projects, in sharp contrast to the temporary situation of war. Connecting the Free Medical School to the history of the Valdecilla Hospital and its successful research and postgraduate teaching also enabled him to show the new regional authorities that there were already favourable conditions and experienced teachers for the establishment of an official Faculty of Medicine at the end of the conflict.[53]

In this respect López Albo was not improvising. Already since the 1920s he had participated actively in debates about the reform of medical teaching, among other things representing neuropsychiatry, which otherwise did not have an academic voice. He consistently advocated education with a solid scientific basis, in contrast to teaching based on the nigh incommunicable power of the clinical gaze.[54] He shared this position with such influential colleagues as Marañón and Trías Pujol. The latter, as Dean of the Faculty of Medicine in Barcelona during its brief period of autonomy, was responsible for reform guided by this criterion. Trías captured the spirit of such reform eloquently in 1935:

---

48 *EC*, 21 November 1936.

49 *LVC*, 24 November 1936; *EDM*, 24 November 1936; *EC*, 24 November 1936.

50 W. López Albo, 'Discurso Inaugural del Instituto Médico de Postgraduados', *Archivos de Medicina, Cirugía y Especialidades* 31 (1930), 40–41.

51 Ibid., 41.

52 A system of fellowships and loans was already put into practice by December 1936. *EC*, 8 December 1936.

53 *EDM*, 24 November 1936.

54 Salmón, García Ballester and Arrizabalaga, *Casa de Salud* (n. 33), 177–214.

> We can say that medicine is always research, both at the moment of interpreting a symptom accurately and establishing a diagnosis, and in the indication of a treatment, as in the discovery of Koch's bacillus or of the endocrine function of the thyroid gland. *Everything is scientific research*. And from the most modest to the most far-reaching, the same mental procedures, the same methods are involved.[55]

It was this orientation, which unified the sickbed and the laboratory, that López Albo had already promoted under the dictatorship of Primo de Rivera as the central idea for a new university 'that would emerge almost automatically in a democratic regime'.[56] And he continued during the early years of the Second Republic:

> A university thus created would be a real crucible of science and culture, in which each intellectual worker would be entitled to time for research.... In this university, profit-driven practice would be minimised and professionals of science would be trained; and the university spirit would not be suffocated by the commercial spirit.[57]

This ideal was shared by many of his colleagues equally critical of traditional medical teaching and, more generally, of the Spanish university during the dictatorship. They saw in the new republican regime an opportunity to complete reform in this direction.[58] Science as a unifying category was crucial in the creation of the new professional, and it is telling that López Albo chose not to speak of physicians but of scientific professionals (*profesionales científicos*) who would be able to understand that science, in the words of Marañón, was 'not a heroic feat, but an attitude'.[59] Undoubtedly this attitude had an ethical and political

---

55 A. Trías Pujol, 'Discusión al Tema "Enseñanza Médica"', *Gaceta Médica Española* 10 (1935), 107-p; our emphasis.

56 W. López Albo, 'El Espíritu Universitario', *El Liberal* (hereafter *EL*), 16 February 1932.

57 W. López Albo, 'Modernas Orientaciones Universitarias (Conclusión)', *EL*, 11 March 1932.

58 See for example the opinions expressed by José Sánchis Banús, Gregorio Marañón and Pio del Rio Hortega in a meeting organised by the progressive Medical Student Association, published in 'Reforma de la Enseñanza Universitaria', *Archivos de Medicina, Cirugía y Especialidades* 33 (1930), 489–502. Rodríguez Lafora, a leading member of the AEN, took up the theme of reform again during the Civil War. See G.R. Lafora, 'Crisis y Futuro de la Universidad', *Cuadernos de la Casa de la Cultura* 1 (1937), 97–104.

59 G. Marañón, 'Reforma de la Enseñanza Universitaria', *Archivos de Medicina, Cirugía y Especialidades* 33 (1930), 500. Reform is a constant theme in Marañón's thought. See, for example, his closing speech at the Valdecilla Hospital's summer courses: idem, 'Concepto de Medicina Nacional', *Gaceta Médica Española* 6 (1931), 18–24-p. A recent update of the historical significance of the figure and work of Gregorio Marañón can be found in a special issue of *Arbor. Ciencia, Pensamiento y Cultura* 189 (794) (2013).

dimension for López Albo, and this is how the medical press understood the work he carried out in the scientific organisation of the Valdecilla Hospital with the creation in 1930 of the Postgraduate Medical Institute, aimed at training specialists. It was to this successful initiative that López Albo referred in his speech of 1936 as the immediate precursor of the Free School of Medicine.[60]

The existing Postgraduate Medical Institute of the Valdecilla Hospital provided the human and material resources for the opening of the Free School. The medical library was the largest in Spain, and every department had a well-equipped laboratory for research. The head of each clinical service also had a contractual commitment to research and teaching, and some of them combined their work in the hospital with a Chair in the Faculty of Medicine of Valladolid. In addition, the staff at the Free School of Medicine included university professors loyal to the republic who, having worked on the summer courses of the Valdecilla Hospital, felt it was safer to remain in Santander during the conflict.

The programme on offer was the centrally prescribed curriculum, although López Albo did not forget the discussions on reform of medical teaching. The proposed new syllabus was structured around physiology linked to pathology in a single category alongside preventive medicine. The reform discussions had crystallised in proposals of the First Congress for the Advancement of Medicine in 1935, some of which were formalised in a decree on medical teaching in February 1936.[61] In his inaugural speech of the Free School, López Albo stressed the need to incorporate the new scientific approach, uniting practice with theory and making disciplinary boundaries unnecessary: '[T]he Valdecilla Hospital reserves the right to prepare for the next course a plan with its own orientation, based on the practical and integrated teaching of medicine'. He added confidently:

> Our school will not in any way be a factory of physicians. Its norm will be quality, not quantity. It will have a *numerus clausus*, achieved by rigorous selection of access to the School and taking into account the real teaching capacity of the institution. We believe that this figure must be between 150 and 200 students. The Valdecilla Hospital, after its first two accomplishments (the Postgraduate Medical Institute and the Nurses' School), today starts out on a new teaching route. And to achieve this will require no special effort.[62]

Not only were his words greatly optimistic, trusting in a rapid resolution of the war, they also reinforced, importantly, the idea of continuity of a scientific

---

60 *Gaceta Médica Española*, 4, 1930.

61 The lectures and discussions of the conference and a detailed analysis of the decree were published in *Gaceta Médica Española* 10 (1935–36).

62 *EDM*, 24 November 1936.

and social project that, begun during the dictatorship of Primo de Rivera, was genuinely republican from the start. Science, education and social commitment seemed necessarily linked to the republican ideology. And when López Albo recalled with gratitude the Marqués de Valdecilla who had died in 1932, he played on the way in which his figure had been transformed. In the popular imagination of the 1920s the marquis had been a paternalistic secular saint, now, in the Civil War, he was a working class hero. For the *Frente Popular* (Popular Front), Ramón Pelayo was a rich philanthropist who, engaged in educational and socio-scientific works for the benefit of those 'who have nothing', never forgot his working class roots. In their words, he represented 'the ancient nobility of work, which is far superior to that of aristocratic titles'.[63] And it was precisely in Pelayo's mansion 'La Cabaña' ('the hut') that the farm-asylum for the mentally ill was established in January 1937.[64]

The Ramón Pelayo Farm-Asylum

The province of Santander did not have its own infrastructure to care for chronically ill mental patients, and if these patients did not have the means to pay for a private sanatorium they were transferred to the charitable mental asylums of Valladolid and Palencia. The Civil War made this transfer impossible and, from the autumn of 1936, the Valdecilla Hospital was obliged to admit such patients into its psychiatric pavilion. The situation was undoubtedly bad news for López Albo, recently reappointed as director of the hospital. From the time he designed the psychiatric service in 1929, he had insisted that this was a unit exclusively for acute cases. But neither his displeasure nor the clinical problems that these new patients posed were sufficient reason for the diligence with which he dedicated himself to trying to solve the problem. As we shall show, there was a bigger plan at work that had been in the making for over a decade.

In barely four months the former residence of the Marqués de Valdecilla near Santander was refurbished into a farm-asylum, which opened its doors on 31 January 1937. Its direct function was to admit the new influx of patients. But it was also meant to apply various therapeutic techniques that López Albo had learned first-hand from the principal European insane asylums during a study trip.[65] This trip had been commissioned by the Provincial Assembly of the Basque country and the whole farm-asylum project was part of a regional plan for psychiatric organisation.

---

[63] *EC*, 24 November 1936.

[64] *EC*, 2 February 1937; *LVC*, 2 February 1937.

[65] W. López Albo, 'El Sanatorio Psiquiátrico de Zamudio y la Escuela-Granja Agrícola', *EL*, 26 February 1932.

This regional plan was a continuation of a personal project of López Albo, which he had already drafted in 1926, failed to put into practice fully in Santander in 1929 and articulated for the Basque country in 1935.[66] It represented the trend of majority opinion of his colleagues in the AEN and the League of Mental Hygiene, which they had expressed on numerous occasions. In terms of concrete political measures, the 1935 Draft Law for Organisation of National Psychiatric Care had come closest to realise the ideas of López Albo and the AEN. It was ambitious in terms of social intervention but went in abeyance because of the war.[67] In Santander, the psychiatric pavilion of the Valdecilla Hospital brought some realisation. Following López Albo's design, it had been built in two large centres: the dispensary, which included the surgery and social welfare services, and the mental clinic.[68] The incorporation of psychiatric services in a general hospital was unusual in Spain and no doubt brought professional benefits to its founders. At the same time, it reflected a new social awareness and scientific approach.[69]

The dispensary – whether associated with a general hospital or not – was one of the pillars of psychiatric reform promoted by the AEN. The introduction of early discharge for the chronically ill and the creation of open clinics for the acutely ill required an infrastructure for extramural care – a subject that occupied many pages in the professional and national press in the 1920s.[70] The proposals of José Manuel Sacristán and Oscar Torras are well-known examples, as are those collected in the Proceedings of the Second National Meeting of the League of Mental Hygiene held in Bilbao in 1928.[71] It was during this meeting that López

---

66 López Albo, 'Plan Moderno' (n. 33); idem, 'El Problema de los Enfermos Psíquicos en Vizcaya. Su Solución Eficiente y Económica', *Revista Clínica de Bilbao* 10 (1935), 34–7.

67 Huertas 'Organizar y Persuadir' (n. 6), 208–10.

68 W. López Albo, 'Proyecto de un Pabellón Psiquiátrico, Dispensario Psiquiátrico e Instituto de Orientación Profesional para la "Casa de Salud Valdecilla"', *Revista Clínica de Bilbao* 3 (1928), 581–6.

69 J.M. Sacristán, J. Germain, and B. Rodriguez Arias, 'Psiquiatría Práctica e Higiene Mental (Impresiones de un Viaje a los Estados Unidos)', *Archivos de Neurobiología* 10 (1930), 372. J. M. Comelles, *La Razón y la Sinrazón. Asistencia Psiquiátrica y Desarrollo del Estado en la España Contemporánea* (Barcelona, 1988), 103–51.

70 R. Huertas, 'Mental Health and Psychiatric Care in the Second Spanish Republic', *History of Psychiatry* 9 (1998), 51–64; idem, 'Organizar y Persuadir' (n. 6), 202–8.

71 J.M. Sacristán, 'Para la Reforma de la Asistencia a los Enfermos Mentales en España', *Archivos de Neurobiología* 2 (1921), 1–15; idem, 'Plan Moderno de Asistencia a los Alienados', *Archivos de Neurobiología* 6 (1926), 226–31; O. Torras, 'Las Modernas Orientaciones de la Asistencia a los Enfermos Mentales', *Psiquiatría* 2 (1923), 21–7; idem, 'Los Servicios Libres Psiquiátricos', *Psiquiatría* 2 (1923), 20–29; idem, 'Los Dispensarios Psiquiátricos y los Servicios Sociales de Asistencia', *Psiquiatría* 2 (1923), 22–9.

Albo presented his 'Project for a psychiatric pavilion, psychiatric dispensary and institute of career advice' for the Valdecilla Hospital.[72] But it was not until the decree on care for the mentally ill of July 1931 that it became obligatory to set up these dispensaries.[73]

López Albo conceived the psychiatric dispensary of the Valdecilla Hospital along these lines but targeted it at a wider public – not just mental patients. It offered two distinct services: a psychiatric surgery and a social welfare service that would keep records of the social and economic conditions of the patient. The latter was also in charge of a whole range of wider functions: the prevention of antisocial acts, interventions in the family and neighbourhood setting, campaigns of prevention and mental hygiene, the reintegration of the patient in society, the promotion of care associations for the mentally ill and the channelling of welfare resources.[74] Moreover, the service would be complemented with a career advice institute (*Instituto de Orientación Profesional*) that aimed to place 'each citizen in the craft and job best suited to his physical, sensory, psychomotoric and mental aptitude, so that he can produce the most and the best without fatigue'.[75] López Albo's ideas in this regard stemmed from his interest in Taylorism in the 1920s and similar work in the Soviet Union in the 1930s.[76]

In reality what was being planned was a social organisation governed by scientific principles strongly linked to measures of social hygiene and eugenics combined with a particular reading of the science of work.[77] However, to transfer these proposals of social intervention to the Spanish reality of the 1930s was nigh impossible given the poor administrative infrastructure of the state. In fact, despite the legal requirement of creating psychiatric dispensaries,

---

72 Anon., 'Liga Española de Higiene Mental. (II Asamblea. 21, 22 y 23 de Septiembre. Bilbao)', *Archivos de Neurobiología* 8 (1928), 293–4.

73 'Asistencia de Enfermos Psiquiátricos', decree of 3 July 1931 (*Gaceta*, 7 July 1931).

74 Salmón, García Ballester and Arrizabalaga, 'Casa de Salud' (n. 33), 169–77.

75 W. López Albo, 'Medicina Social. La Orientación en los Oficios. Selección de los Obreros y Determinación de sus Aptitudes Profesionales', *Revista Clínica de Bilbao* 2 (1927), 97–8.

76 Salmón, 'Wenceslao López Albo' (n. 36), 323–59. López Albo learned Russian in 1934 and strongly recommended its mastery to medical students. See W. López Albo, *El Médico y los Idiomas* (Bilbao, 1934).

77 On the development of the science of work in Spain under Primo de Rivera's dictatorship and the Second Republic, see J. Martínez Pérez, 'La Organización Científica del Trabajo y las Estrategias Médicas de Seguridad Laboral en España (1922–1936)', *Dynamis* 14 (1994), 131–58. See also R.M. Medina, E. Rodríguez Ocaña, 'La Medicina en la Organización Científica del Trabajo. El Instituto de Orientación Profesional (Psicotécnico) de Barcelona (1917–1936)', in R. Huertas and R. Campos (eds), *Medicina Social y Clase Obrera en España (siglos XIX y XX)* (Madrid, 1992), vol. 2, 459–90.

only one was set up during the first two years of republican government.[78] A similar gap between plans and reality can be seen in the Valdecilla example: in practice, neither the social welfare service nor the career advice institute became operational. The psychiatric surgery did function from the opening of the Valdecilla Hospital in 1929, as did the mental clinic. The organisation of the mental clinic was similar to that of the other specialised clinics in the hospital, except for equipment needed specifically for these patients.[79] López Albo organised it in such a way that it could not become an insane asylum in miniature. For this reason, he took special care in formulating the regulations of the mental clinic, especially in setting limits to the duration of hospitalisation.[80]

In 1936 López Albo combined his aims with economic arguments regarding the overcrowding of the psychiatric pavilion of the Valdecilla Hospital with patients who could not be transferred to Valladolid and Palencia anymore. Thus he convinced the Board of the hospital and the provincial assembly to give him the opportunity of creating a farm-asylum as part of a far-reaching social plan. What this plan entailed, he explained in a two articles in *El Cantábrico* on 28 and 29 January 1937. In several stages, and according to the most up to date scientific guidance, the entire province would be covered with farmhouse schools, farm-asylums, extramural healthcare in rural environments, a system of dispensaries in urban settings and centres for career advice – all with the psychiatric pavilion of the Valdecilla Hospital as the controlling body.[81] This plan would be completed with a modern social welfare department which would be distinguished 'for its scientific character' and from the traditional charity and philanthropy typical of feudal and capitalist systems. According to López Albo, after the republican triumph in the war, the sense of social solidarity would grow, and social welfare would be a public duty. This would lead to a scientific and humanitarian organisation of society:

> Until recently, Spain has remained in an almost embryonic state with regard to these issues. It is hoped that the Republic, when it emerges strengthened from the present misfortunes, will give healthy impetus to the dynamism of society and thus stimulate and awaken us from the lethargy in which we were plunged before the change of regime.[82]

---

78 Ministerio de la Gobernación, *La Sanidad en la República* (Madrid, 1933), 17.

79 W. López Albo, 'Consultas de Neurología, Neurocirugía y Psiquiatría', *EDM*, 24 October 1929.

80 López Albo, 'Proyecto de un Pabellón' (n. 68), 585–6.

81 W. López Albo, 'El Problema de los Enfermos Mentales en Santander. Su Solución Eficiente y Económica', *EC*, 28 and 29 January 1937.

82 W. López Albo, 'Orientaciones Modernas en Asistencia Social', *EC*, 14 February 1937.

Hence, the project was not restricted to the health of mental patients, but also aspired to build the foundations of a republic designed scientifically. Under the umbrella of the prevention of mental maladjustment, the plan envisaged the regulation of all aspects of social life on scientific principles. The individual would be monitored even antenatally, and psychology and pedagogy would guarantee a new social order where access to education and positions of social responsibility would only be limited by individual capacities and not by class differences. In this regard it is revealing that, a few months before the war, López Albo complained that the new decree on medical education omitted the proposals of the well-known Catalan psychologist Emili Mira that psychological examination should take priority over school examination results in the recruitment of medical students.[83]

Prevention was the key word in a plan of dubious theoretical consistency that borrowed from various sources, from eugenics to the science of work, especially as developed in the Soviet Union. The rationality of science would provide the theoretical and practical elements to prevent almost all ailments (physical, psychological, social as well as political) that the new Second Republic had just started to combat when fascism put a halt to its potential. López Albo never fully explained how the society he envisaged would be organised on scientific principles, but it seemed to be ruled by a self-organising principle that would be put into motion almost automatically when the minds of the people, liberated from the claws of religion, were enlightened with the rationality of the scientific method.[84] More generally, the frequent use of 'science' and 'scientific' merged with the style of the utopian literature advocating social intervention, both of a reforming and of a revolutionary character, that was current in the 1930s.[85] Under the circumstances imposed by the war, the prospect of a scientifically designed social order acquired new significance as a comforting promise amidst uncertainty and insecurity.

The confiscation of the assets of the heirs of the Marqués de Valdecilla provided the material infrastructure, and López Albo provided the scientific argument for the farm-asylum for chronically ill mental patients.[86] The asylum was inaugurated as the '*Colonia Agrícola Ramón Pelayo*' (Ramón Pelayo Farm-Asylum) and by May 1937 already gave shelter to 54 patients segregated according to sex in two buildings (Figure 7.4).[87] At the same time, López Albo

---

83 E. Mira, 'Enseñanza Médica. Métodos de Selección de los Alumnos', *Gaceta Médica Española* 10 (1935), 48–9 and 73–6. See also W. López Albo, 'Algunos Comentarios al Nuevo Plan de Eenseñanza Médica', *Gaceta Médica Española* 10 (1936), 135–8-p.

84 See especially López Albo, 'Espíritu Universitario' (n. 56); idem, 'Modernas Orientaciones' (n. 57) and idem, 'Orientaciones Modernas' (n. 82).

85 For example, G. García Maroto, *La Nueva España,* 1930 (Madrid, 1927).

86 López Albo, 'Problema de los Enfermos' (n. 81).

87 *EC*, 2 February 1937; *LVC*, 2 February 1937. *LVC*, 20 May 1937.

REDACCIÓN, ADMINISTRACIÓN Y TALLERES, CALLE DE MARCOS UNAZASORO, NÚM. 19. APARTADO POSTAL 62. TELÉFONO 15-55

LA VOZ DE CANTABRIA

MANTENEMOS COMUNICACION DIRECTA CON NUESTRA AG... MADRID, HASTA LAS 4 DE LA MA...

LA COLONIA AGRICOLA "RAMON PELAYO"

PROVECHOSA LABOR MEDICA.-EL FUNCIONAMIENTO DE UN MANICOMIO PROVINCIAL.-UN PROYECTO QUE DEBE SER TOMADO EN CONSIDERACION POR TODOS LOS ORGANISMOS OFICIALES

INFORMACION LO...

Bajo la presidencia del ministro Agricultura, señor Uribe, tuvo lu... ayer en la Delegación del Gobier... una interesante reunión, a la que a... tieron diversas autoridades de las ... provincias del Norte

Figure 7.4 Article about the functioning of the Ramón Pelayo Farm-Asylum, from *La Voz de Cantabria,* 18 April 1937. The top photograph shows the staff with, in the front row, López Albo in the centre and a patient on the left. The bottom photograph shows patients working on the farm.

Source: Reproduced with permission from the Biblioteca Municipal, Santander.

did clinical research in the psychiatric pavilion of the Valdecilla Hospital and offered a two-year course of psychiatric specialisation for qualified nurses.[88]

The farm-asylum followed the structure developed in various centres in Germany and Holland and applied the principles of occupational therapy developed by Hermann Simon.[89] This was of great economic as well as

88 The research results were published in W. López Albo, 'La Aglucorraquia en el Diagnóstico de la Cisticercosis del Sistema Nervioso Central. (Quinto Caso Diagnosticado en Vida, en España, y Primero Verificado)', *Revista de Sanidad de Guerra* 1 (1937), 275–82. A handbook for the course was published by E. Pelaz Martínez, *Asistencia al Enfermo Mental* (Santander, 1937).

89 López Albo, 'El Sanatorio Psiquiátrico' (n. 65); H. Simon, 'Defensa de la Terapéutica Ocupacional', *Archivos de Neurobiología* 26 (1936), 575–82.

scientific importance, allowing food self-sufficiency in wartime. Setting mental patients to work in the farm instead of letting them languish in ordinary asylums also meant that these patients were given a valuable social role. Hence, the farm-asylum incorporated the latest science-based psychiatric practices in conjunction with a vision of a fairer, more unified society from which the chronically ill mental patient was not excluded.

The message was clear: continuity and confidence in a republic that, strengthened after the struggle against fascism, could develop the potentialities that its short life had impeded. Science would play a key role, and López Albo's projects were designed to show that they would outlive any sense of provisionality because of the war. The temporary barbarity would be superseded by the long-term rationality of science. The Valdecilla Hospital did not offer a fast-track course on psychiatric nursing, such as those offered on general and orthopaedic surgery of war, but a two-year specialist education. The overcrowding of the psychiatric pavilion was not solved with a depot for chronic mental patients, but with a complete plan of social engineering where psychiatry would play an organising role in forging new social relations.[90] Thus, López Albo took advantage of the war circumstances by finally putting into practice the proposals debated by the members of the AEN and the League of Mental Hygiene since the 1920s, where prevention of mental maladjustment was paired with pedagogical intervention and eugenic measures.

Similarly, continuing research and laboratory routines helped to sustain confidence in the republican project despite the alarming course of the war. López Albo's research on brain parasites is a good example. After fleeing Santander, he wrote: 'Among the sick and wounded who passed through our clinics of the Valdecilla Hospital during eleven months of the war (from September 1936 to August 1937), there is one interesting case of encephalic compression due to cysticercosis'.[91] The article was as rigorous as those he published before, but it was incomplete because of the adverse circumstances: 'Having been forced to flee Santander when it was invaded by the Italian and Falangist troops, we are prevented from completing this work with pathological-anatomical data'.[92] López Albo was not deaf to the bombing, and the invasion cannot have taken him by surprise. But he believed strongly that setting an example by continuing his research was the best way to defend the values of rationality that he considered to be integral to the republican project. Other colleagues from the AEN took a similar path.[93]

---

90 López Albo, 'Orientaciones Modernas' (n. 82).

91 López Albo, 'La Aglucorraquia' (n. 88), 275.

92 Ibid., 280, n. 1.

93 Huertas, *Organizar y Persuadir* (n. 6), 112; J. Lázaro, 'La Significación de *Archivos de Neurobiología* en la Psiquiatría Española de 1936', *Archivos de Neurobiología* (Supplement 1) (1997), V–XIX.

Numerous articles appeared in the general press between February and April 1937 describing the peaceful life of the patients in the farm-asylum – in contrast to the fascist bombardments reported from the Basque country. A headline in *La Voz de Cantabria* of April 1937 reads, 'Healthy Living and Food, Medical Attention and Relative Freedom Offer the Mentally Handicapped of Cantabria a Future Far Removed from the Dramatic Shadow of Irrationality'.[94] Whose irrationality? Not that of the mental patients, but of the deranged fascists. The Ramón Pelayo Farm-Asylum became a tragic metaphor for a republic to which scientists were committed as a project of reason. There was no possibility of entering into dialogue with the irrational.

Conclusions

War is about destruction, not only material but also symbolic. This was the case in the 1930s, as it is now. War could also be experienced, however, as a space for creative possibilities within a confused framework of mixed spontaneity and coercion. This was even more so in the Spanish Civil War where the war was waged against a short-lived new social and political experiment, the Second Republic, which was just beginning to explore its potential. Science was an important feature in the rhetoric that furnished the definition of the republican project. In this context science was much more than a circumscribed set of practices and the motor of technical and material progress. Science was a general attitude in which the secular values of free enquiry and objectivity were combined with a commitment to the construction of a fairer new society. The AEN not only embodied those values, but was also given the opportunity to bring them to fruition in particular projects where the same methodological procedures were advocated for neurohistological research and for preventing social maladjustment. Thus, it seems only logical that a member of the AEN, López Albo, saw the outbreak of the war not as a closing barrier but as a point of departure for exploring further the possibilities of science as a guiding principle of social organisation.

The line, 'I am your choice, your decision', from W.H. Auden's poem 'Spain' of 1937, eloquently captured the Spanish Civil War as an open space where the most contradictory interventions were possible. López Albo resolutely made these opportunities his own. The Free School of Medicine and the farm-asylum were created, not as short-term projects to solve particular problems caused by the circumstances, but as far-reaching programmes with a wider potential stretching into the future. The Free School aimed to change both medical teaching and practice by projecting science as a unifier of the sickbed and the laboratory, practice and theory, medicine and surgery, and pre-clinical

94 *LVC*, 21 April 1937.

and clinical mentalities – all integrated in a new coherent syllabus. The farm-asylum was not designed as a temporary solution to the problem of an overcrowded hospital, but as part of a science-driven society. Both projects fused science and republic and served as powerful unifying icons amidst uncertainty and confusion.

The Free School of Medicine and the farm-asylum were dismantled after the entry of Franco's troops into Santander in August 1937. All institutional records were made to disappear in an attempt to erase both initiatives from memory.

Acknowledgements

We would like to thank Harmke Kamminga and Geert Somsen for their careful reading of the paper and stimulating comments. The research has been funded by the Spanish Secretary of Economy and Competitiveness (HAR2012-37754-C02-01).

Chapter 8

'To formulate a plan for better living'

Visual communication and scientific planning in Paul Rotha's documentary films, 1935–45

Timothy Boon

This chapter argues that the visual articulation of scientific and technical modernity was integral to endeavours to unify science and society.[1] It takes as its main focus Paul Rotha (1907–84), the British film-maker who most took to heart issues of planning and the social role of science. By the outbreak of the Second World War, Rotha had established himself both as a major documentary director, with 20 films to his name, and as a respected author of four substantial books and numerous articles advocating documentary in the context of film criticism.[2]

I will show that Rotha's interest in planning and science was stimulated by contacts he made in the 1930s with groups of architects, scientists and journalists, most of them on the political left. The films made by Rotha between the mid-1930s and the end of the Second World War illustrate the translation into the public sphere of private discussions within these groups on themes of planning, the social usefulness of science and their visual communication.[3] The story culminates in the collaboration between Rotha and Otto Neurath (1882–1945) in the production of documentary films between 1941 and 1945.

Neurath also had a strong commitment to using visual communication techniques to promote scientific planning to the general public. Specifically, he developed Isotype, the universal picture language, which he used to convey scientific and statistical information in publications and exhibitions. Neurath's and Rotha's

---

1 This account extracts one of the threads of argument also to be found in T. Boon, *Films of Fact: A History of Science in Documentary Films and Television* (London, 2008).

2 For fuller biographical and bibliographical information, see T. Boon, '"The Shell of a Prosperous Age": History, Landscape and the Modern in Paul Rotha's *The Face of Britain* (1935)', in C. Lawrence and A. Mayer (eds), *Regenerating England: Science, Medicine and Culture in the Interwar Years* (Amsterdam, 2000), 107–48, esp. 107–11; and D. Petrie and R. Kruger (eds), *A Rotha Reader* (Exeter, 1999).

3 A useful summary of social, economic and town planning can be found in J. Stevenson, 'Planner's Moon? The Second World War and the Planning Movement', in H.L. Smith (ed.), *War and Social Change* (Manchester, 1986), 58–77.

shared stress on visual communication was joined by a common commitment to communicate to the general public the transformative power of science and planning. As I will show, the collaboration between them brought innovation for both: using Isotypes in his films gave Rotha a new visual means to impart factual information, whilst it gave Neurath the opportunity of broadening his repertoire of visual communication to films. The earlier sections of this chapter, by showing how Rotha became an advocate of scientific planning, demonstrate a broader thesis that Neurath's views fell on prepared ground in Britain.

Paul Rotha in the 1930s: documentary, architecture and science

Even before the six months he spent at the Empire Marketing Board (EMB) Film Unit under John Grierson in 1930, Rotha had made his mark in the world of cinema with his book *The Film Till Now*, which he published in 1929 at the age of 22.[4] For much of the 1930s Rotha made films at Gaumont-British Instructional, a rather uncomfortable niche, as his assumptions about film-making differed markedly from those of his colleagues. His 1930s films included *Contact* (1933, which treated international aircraft travel), *Shipyard* (1935, on the social relations of the construction of a liner) and *The Face of Britain* (1935, which is discussed below). But his significance rests as much on his prolific written work as on his films. He was composing *Documentary Film*, the first book-length justification of the genre, at the same time as he was making *The Face of Britain*.

Working away from the main concentrations of documentarists (at the EMB and later General Post Office Film Units) seems to have made network-building especially important for Rotha. Certainly, it was through more and less formal networks that he produced most of his films. By following some of Rotha's contacts, we can trace the development of a set of shared concerns that eventually produced films such as *World of Plenty* (1943) and *Land of Promise* (1945), in which Rotha produced forceful expressions of the virtues of a scientifically organised and planned society by fusing the visual rhetoric of Isotype with the formalism of Eisensteinian cinematic editing technique.

Of Rotha's work from the 1930s, *The Face of Britain*, the fifth of his documentaries, best exemplifies his emerging commitment to modernist causes and to scientific planning. Rotha intended this film as a self-conscious document of the state of the nation in the mid-1930s. He described it as 'a film of the natural and scientific planning of Britain with reference to the respective power of coal and electricity'.[5]

---

4 See P. Swann, *The British Documentary Film Movement 1926–1946* (Cambridge, 1989).

5 P. Rotha, *Documentary Film* (3rd edn, London, 1952), 255.

Figure 8.1 'The New Power'.
Source: From *The Face of Britain* (British Film Institute).

The Face of Britain is structured as a historical dialectic: 'Heritage of the past', the first of four historical sections and analytically the film's thesis, paints an idyllic view of Britain's community and landscape before industrialisation. 'The smoke age' provides the antithesis, describing the catastrophic impact of the industrial revolution on the people and land of Britain. The synthesis is achieved jointly by 'the new power' (Figure 8.1), which introduces hydroelectric generation, and 'the new age', which demonstrates the planning of Britain that, the film argues, electrification permits. The commentary of this final section, the culminating synthesis of the film, is vibrantly modernist:

> Strange new architectures arise to meet the demands of a changing civilisation, shapes and forms of simple beauty, dictated only by the purpose which they are meant to serve. Out of steel, glass and concrete, the architects and engineers must transform the face of Britain. . . . New sources of power, new means of communication, new methods and new processes are here for the service of man.

In a hearty espousal of scientific planning that was novel for Rotha's work in 1935, the commentary continued, 'this is an age of scientific planning, organisation of cooperation and collective working. . . . Machinery and science can bring about better conditions of working and permit greater freedom for leisure'.

Rotha adopted the dialectical montage of Russian film theorist and director Sergei Eisenstein, who, together with other Soviet film-makers such as Pudovkin, used dialectical materialism as a model to structure films. Rotha's commitment stemmed from his belief that the Russians had discovered the 'correct' way to make films. He was amongst those who believed Russian montage theory to be *the* rigorous and scientific approach to film construction. Ivor Montagu's preface to the English edition of Vsevolod Pudovkin's *Film Technique* compared its impact on film to that of Mendel's genetics on plant breeding.[6] In a review of this book, Rotha stated that most (Hollywood) story films had been produced 'upon basically wrong principles'.[7] Much later he wrote to Pudovkin:

> I should like to say, my friend, that we documentary workers in England . . . shall always acknowledge our great debt to your film technicians of the Soviet Union. It was your great films, and those of Dovzhenko and Eisenstein, in those far-off silent film days, from which we learnt so much both in concept and technical handling.[8]

Part of the claim of Russian montage to being 'scientific' and 'true' lay in the experiments carried out by people such as Lev Kuleshov into how editing created and conveyed meaning. An example is what became known as the 'Kuleshov effect' in which, by editing together two different shots, a third meaning was shown to be produced.[9]

The Face of Britain could be said to straddle the fence of modernity and modernism; it represents modernity – and the history of its arrival via industrialisation – and it uses modernistic technique to do this. The film might, in this double sense, have been seen by contemporaries as scientific: first, it argued that science, technology and architectural modernism provide the means to ameliorate the problems created by industrialisation. Second, the use of 'scientific' Russian montage theory was chosen to convey it. At the same time, it is clear that Rotha was more concerned to adopt a method that made good *cinematic* sense than to conduct an argument in the fine detail of Marxist theory.[10]

6 V. Pudovkin, *Pudovkin on Film Technique: Five Essays and Two Addresses*, trans. I. Montagu (London, 1933).

7 P. Rotha, 'Pudovkin on Film Technique (review)', *Sight and Sound* 1 (1933), 140.

8 Rotha to Pudovkin, 24 February 1947, folder 5, Box 66, Rotha Archive, Department of Special Collections, Charles E. Young Research Library, University of California at Los Angeles (hereafter 'Rotha Archive'). This was Rotha's reply to a letter he received from Pudovkin in praise of *World of Plenty*, which will be discussed later.

9 R. Taylor and I. Christie (eds), *Inside the Film Factory* (London, 1991), 44–5.

10 This is explored in Boon, 'Shell of a Prosperous Age' (n. 1). See also P. Rotha, *Documentary Film* (London, 1936), 117–18.

Like so many on the left, Rotha was the member of no party; in fact he responded with some indignation to requests to reveal his party affiliation: 'I tied myself to no party other than the human race. I am no fly or bug to be catalogued or labelled for all mankind to expect me to live by a written doctrine'.[11] Still, his left-wing stance was hardly hidden. When *The Times* stated that 'the aesthetic left in the cinema is also ranged on the left politically', they were taking Rotha as a prime example.[12] Indeed, his interest in dialectical montage and in planning, both central to *The Face of Britain*, were typical concerns of those on the left. Furthermore, some may well have read the portrayal of electricity as a social transformative force as a reference to Lenin's equation that 'communism is Soviet power plus the electrification of the whole country'.[13]

The *Face of Britain* owed much, not only to Russian inspiration, but also to the influence of Rotha's key associates at this time, especially as regards its theme of planning. Hugh Quigley, Chief Statistical Officer of the Central Electricity Board (CEB) since 1931, was the most significant sponsor and contact. Quigley had already proclaimed the virtues of planning even before he joined the CEB in his volume *Electrical Power and National Progress* (1925). And, in the depressed 1930s, he propagated a rationalist planning response to the social problems of housing.[14] Rotha later recalled, on meeting Quigley: 'We at once saw eye to eye about a film on the possibilities of replanning Britain, both industrial and rural, based on the flexible power of the national grid'.[15]

Rotha's interest in planning and in the changing English countryside was also strongly influenced by his contacts with the architectural profession. In his autobiography he recalled that articles in the *Architectural Review* had been one important source of inspiration for *The Face of Britain*. Among architects, Basil Ward was a close ally from the early 1930s. Ward was a partner in the important modern movement architectural practice Connell, Ward and Lucas, which is most

---

11 Rotha to Knight, 19 July 1941, 3, Box 26, Rotha's letters to Eric Knight, Rotha Archive.

12 Anon., 'The Cinema and Society: Documentary Films (Review of Rotha's *Documentary Film*)', *The Times*, 4 February 1936.

13 This slogan had been popularised in Britain at the International Congress of the History of Science and Technology, in 1931; see especially M. Rubinstein, 'Electrification as the Basis of the Technological Reconstruction of the Soviet Union', in G. Werskey (ed.), *Science at the Cross Roads: Papers Presented to the International Congress of the History of Science and Technology, Held in London from June 29th to July 3rd, 1931, by Delegates of the U.S.S.R.* (London, 1971, 2nd edn), 115–45.

14 H. Quigley, *Electrical Power and National Progress* (London, 1925); H. Quigley and I. Goldie, *Housing and Slum Clearance in London* (London, 1934).

15 P. Rotha, *Documentary Diary: An Informal History of the British Documentary Film, 1928–1939*, (London, 1973), 103.

closely identified with the introduction of continental modernist architecture into Britain.[16] Ward was a member of the advisory panel of Associated Realist Film Producers (ARFP), an organisation founded by Rotha in early January 1936 to bring potential clients together with documentary film-makers operating in the commercial sphere.[17] And it was via Ward that Rotha was invited onto the Royal Institute of British Architects (RIBA) public relations committee in 1936. A memorandum discussed by this committee that recommended the production of a programme of architectural films clearly bears Rotha's stamp, illustrated as it is with stills from *The Face of Britain*.[18] The document proposed that ARFP act 'in an advisory and directing capacity on behalf of RIBA'.

ARFP is of interest in that its membership not only included all the most prominent documentary film-makers, but also listed as advisors the biologists Julian Huxley, J.B.S. Haldane and Lancelot Hogben, along with Grierson, Ward, the designer Edward McKnight Kauffer and the composer Walter Leigh. (Indeed, the cinema trade press announced: 'Plan for Non-Fiction Films: Professors to Act as Advisors'.[19]) Through this group, Rotha connected himself with wider networks joined in promoting an ideology linking science, culture more generally and left politics. Although it is not possible to discern an explicit impact of these scientists on the content of Rotha's films, their involvement in documentary was real, as is indicated by Julian Huxley's presence on the list of lecturers when ARFP launched a lecture service in 1936 on films, their history and production.[20]

---

16 See J. Peto and D. Loveday (eds), *Modern Britain 1929–1939* (London, 1999), 19 and 128.

17 Anon., 'The Documentary Film. Consultative Body Formed', *The Times*, 10 January 1936, and P. Rotha, letter to the editor, *Sight and Sound* 8 (1939), 81.

18 Anon., 'Memorandum for the production of a programme of documentary and instructional films on architecture, submitted to the 8 May 1936 meeting of the RIBA Public Relations Committee', vol. 3, Public Relations Committee Minutes (RIBA PUBREL), RIBA Library Manuscripts and Archives Collections, London. In fact, these plans came to nothing due to lack of funds. See T. Haggith, '"Castles in the Air": British Film and the Reconstruction of the Built Environment, 1939–51' (PhD diss., Warwick University, 1998), 46–52.

19 *Kine Weekly*, 16 January 1936, 22. For further details, see T. Boon, '*The Smoke Menace*: Cinema, Sponsorship, and the Social Relations of Science in 1937', in M. Shortland (ed.), *Science and Nature* (Oxford, 1993), 75. I have found no documents that explain how Huxley, Haldane and Hogben were chosen as advisors to the ARFP, but it seems plausible that Ritchie Calder (see below) talked positively about these scientists with Rotha. In *Documentary Diary* (n. 15), Rotha records that the 'scrupulously kept' minute books of the ARFP were mislaid after the war (on p. 161).

20 Anon., 'Syllabus of Lectures on the Film Prepared and Issued by ARFP, 1937–8', folder 2, Box 80, Rotha Archive. See also *The Times*, 30 August 1937, 8.

Rotha nursed ambitions to produce a film on architecture and town planning upon reading Lewis Mumford's *The Culture of Cities* (1938) soon after it came out. He said it 'will tell how Everytown, 1938, England, grew up; how housing and public services have been related, or in so many cases unrelated, to the needs of the community. It could be something for which *The Face of Britain* was a child's notebook'.[21] *Today We Live* (1937), produced by Rotha for the National Council for Social Service, had already examined themes of community in Depression-era Britain.[22] Rotha's 1938 film *New Worlds for Old*, based on reports from the centrist planning organisation Political and Economic Planning (PEP), treated gas and electricity, the utilities vying to dominate the market for domestic fuel supply.

The contact who most stimulated Rotha's ideas about the social role of science was Peter Ritchie Calder, with whom he collaborated from the mid-1930s and through whom he later extended his contacts with circles debating this issue. Calder, science correspondent of the Labour-supporting *Daily Herald* since 1930, had established himself during the 1930s as a respected journalist specialising in science, social conditions and their combination. Although Rotha later recalled that it was in 1935 that he and Ritchie Calder 'found a common aim in the promotion of social progress in science and the social progress of the documentary film', in the pre-war period what they really shared was a brand of political radicalism.[23] An early collaboration expressing their concern for social and political causes was *The Peace Film*, a short punchy piece made in March 1936 to argue against rearmament.[24] During 1936, both men also sat on the Labour Party's Film Committee, an abortive attempt to encourage the use of film for political purposes, as the Conservative Party had already done successfully for several years.[25]

Calder had consistently argued that scientists should be socially engaged. He took an interest in the affairs of the British Association for the Advancement of

---

21 Rotha to Knight, 20 December 1938, 3, Box 26, Rotha's letters to Eric Knight, Rotha Archive. For a brief account of the fate of this project, see Rotha, *Documentary Diary* (n. 15), 228–9.

22 Ibid., 147–9.

23 P. Rotha, 'Lord RitchieCalder', *The Times*, 6 February 1982, 8. On the so-called Social Relations of Science Movement that Calder was part of, see Gary Werskey, *The Visible College: A Collective Biography of British Scientists and Socialists of the 1930s* (2nd ed., London, 1988) and William McGucken, *Science, Society, and State: The Social Relations of Science Movement in Great Britain, 1931–1947* (Columbus, OH, 1984).

24 Ibid. and Rotha, *Documentary Diary* (n. 15), 165.

25 On the Labour Party's Film Committee, see P. Marris (ed.), *Paul Rotha* (London, 1982), 17 and 47–56. On the Conservative Party films, see R. Low, *Films of Comment and Persuasion of the 1930s* (London, 1979).

Science (BAAS) from at least 1932, when he commented in a newspaper article that 'not a single great social issue had been faced by' them.[26] He supported from the sidelines the resolution placed before the general committee in 1933 requesting 'the Council to consider by what means the Association ... [might] assist towards a better adjustment between advances of science and social progress'. On the establishment of the BAAS's Division for the Social and International Relations of Science in 1938, Calder was made a committee member. This division, set up 'to further the objective study of the social relations of science' was the only section of the BAAS to remain active during the war. It organised a string of conferences, including 'Science and World Order' in 1941 and 'Science and the Public: The Public Understanding of Science' in 1943, in both of which Ritchie Calder was centrally involved. As we shall see, Rotha became deeply committed to the view of a socially useful science that Calder espoused; he filmed sections of the first conference and spoke at the second.

The phoney war and beyond: working on war aims

After the outbreak of war in 1939, consensus in Rotha's circles was consolidated around the issues of war aims and planning for post-war reconstruction. These issues were much debated amidst frustration among scientists, architects and film-makers about their perception of the government's failure to use their expertise for the war effort up to the end of 1940. As science journalist J.G. Crowther recalled, 'the early period of the war was very frustrating for the younger scientists. Many of them were burning to do something, but could find no adequate line of work'.[27] Architects, too, experienced a sense of underemployment and complained of the government's neglect of architects' skills in war organisation.[28]

Rotha, in common with many other documentarists, had a similarly frustrating experience. As the war broke out, he was completing his film on *The Times* newspaper, *The Fourth Estate*, after which work dried up.[29] Since the documentary ideology was predicated on the virtue of the medium as a tool of the democratic

---

[26] W. McGucken, 'The Social Relations of Science: The British Association for the Advancement of Science, 1931–1946', *Proceedings of the American Philosophical Society* 123 (1979), 237–64: 242.

[27] J.G. Crowther, *Fifty Years with Science* (London, 1970), 210.

[28] For example, J.K. Winser, 'Review of *Science in War*', *Architects' Journal*, 15 August 1940, 134–5.

[29] F. Thorpe and N. Pronay, *British Official Films in the Second World War* (Oxford, 1980), 67. In the whole of 1940, Rotha only made a five-minute short about silage, *Mr Borland Thinks Again*.

state, many film-makers expected their role to be enhanced during the war. Instead, there was a rapid succession of short-lived regimes in the Films Division of the newly formed Ministry of Information, and arguments between these officials and the documentarists about what sorts of films should be made became very entrenched.[30] Consequently, despite the fact that planning for war propaganda had been underway since 1938, very few opportunities arose for most documentarists during the first 15 months of war, the period of 'phoney war' and Blitz.

After a series of new appointments to the Films Division in spring 1940, including old documentary allies Jack Beddington and Sidney Bernstein, opportunities for work seemed to improve.[31] In May, Rotha submitted a ten-part 'Report on the production and distribution of documentary films' to the Ministry of Information. Amongst the films he proposed 'to maintain public morale and/or to impart information' was a series of six 'about what we are fighting for'. The document is interesting for what it reveals of the development of Rotha's views. An emphasis on war aims is clear; he complained about the negative tone of 'win the war today and think about the future tomorrow' propaganda. He asserted that 'if national morale is to be maintained, it is important to look into the future. It is necessary to give the public a sense of their privileges and responsibilities as citizens of this democratic State and of what they are fighting to protect'. Yet the focus on planning and the role of science and technology seen in *The Face of Britain* is virtually absent, except for a proposed second series of six films on the impact of war, which includes one on 'how the war has affected Culture and Scientific work'.[32]

Rotha's report did not lead to action, and in September he expressed his frustration, under the pseudonym 'Documentary', in the *New Statesman*:

> By and large, the documentary film movement . . . has received unfortunate handling by the Films Division [of the Ministry of Information]. Its memoranda have been requested and then disregarded, its technicians left for long periods of unemployment, and more than once it was asked to perform impossible jobs by Films Division officials unacquainted with film procedure.[33]

During the long months of enforced inactivity in 1940 when no films were being produced, Rotha allied himself to several interlinked groups that were

30 Swann, *British Documentary Film Movement* (n. 4), 150 and 153–60.

31 Ibid., 154–5.

32 'Report on the Production and Distribution of Documentary Films for the Purpose of Propaganda and Information by the Ministry of Information, Part II, Film Production Programme for Great Britain', 5–6, Box 69/10, Rotha Archive.

33 'Documentary', 'The Film at War', *New Statesman and Nation*, 7 September 1940, 230–31. Rotha's diary, 6 September 1940, folder 16, Box 23, Rotha Archive (hereafter, 'Rotha diary').

beginning to discuss post-war reconstruction. This issue soon became central to Rotha's plans for new films.

The purpose of the war beyond the defeat of Hitler received scant official recognition until Churchill and Roosevelt signed the Atlantic Charter between Britain and America on 14 August 1941.[34] As a result of this seeming official lack of interest, the discussion of war aims and of planning for reconstruction became a focus for those who opposed aspects of how the war was being waged. An early meeting point was H.G. Wells's call for discussion of the 'rights of man', in a correspondence in *The Times* that started the day after war broke out.[35] In his letter of 26 September 1939, Wells intended to provoke discussion about what the war was being fought for, although it soon merged with his broader concerns with a world state. Ritchie Calder was among those who responded to Wells's letter and, in the first months of 1940, became secretary to the committee later chaired by former Lord Chancellor Sankey that drafted the outcome of the initiative, Wells's *New Declaration of the Rights of Man*. The committee also included nutrition scientist John Boyd Orr and Richard Gregory, editor of *Nature* and President of the BAAS at the time.[36] Calder arranged substantial coverage of the debate in the *New Statesman* and the *Daily Herald*, and he and Rotha discussed the possibility of making a film on the subject.[37] Although this film did not materialise, the enterprise cemented relationships between several figures in Rotha's and Calder's circle.

Rotha also became involved with the 1940 Council, an alliance of architects, planners and scientists interested in war aims in the sense that they saw the social conditions of the war as an opportunity for introducing a more rational, planned society afterwards. It was founded at a meeting under the title 'Problems of Social Environment and the War', convened by the Housing Centre at RIBA on 2 February 1940. The *Architects' Journal* reported that the chief objective of the meeting was the formation of an autonomous council to watch planning, architecture, housing, economics and social services nationally, 'co-ordinating and assisting the work of existing bodies'.[38] One of the key members of this group chaired by Lord Balfour of Burleigh was Edward Carter, librarian at RIBA and with communist sympathies, although he maintained that the 1940

34 See P. Calvocoressi, G. Wint and J. Pritchard, *Total War: The Causes and Courses of the Second World War* (London, 1989), 223 and 353–4.

35 Collected in *Peace and War Aims: A Correspondence from the Times. (Including H.G. Wells' New Declaration of 'The Rights of Man')* (London, 1939).

36 For further information, see D.C. Smith, *H.G. Wells: Desperately Mortal* (New Haven, 1986) 428–49.

37 Calder's Diary, 3 February 1940, Peter Ritchie Calder papers, Dep 370/186, Manuscript Division, National Library of Scotland, Edinburgh.

38 *Architects' Journal*, 1 February 1940, 132 and 8 February 1940, 156–7.

Council was a broad church appealing to 'people from Balfour to Bernal who respond to the same pattern of ideas'. Carter wanted the 1940 Council to concentrate on research and publicity, and he wanted the focus of publicity to be on 'giving the people the facts' because, he argued, 'schemes for planning must grow out of the people's awareness of the need for improvement of their conditions . . . until the people are aware, they cannot delegate their authority to the experts to work things out for them'. He sought to extend the Council's representation across the sciences, especially the social sciences.[39] One of the most visible of the 1940 Council's activities was the 'Living in Cities' exhibition, produced in collaboration with the British Institute of Adult Education.[40] The 1940 Council was encouraged by the appointment of John Reith as head of the new Ministry of Works in October 1940, with a specific brief to address issues of reconstruction.[41]

Rotha first had contact with the 1940 Council group early in October 1940 when he telephoned Carter following a chance meeting with the architect Judith Ledeboer, a member of the group. He immediately saw the potential of making a 1940 Council film for a return to documentary's purpose of 'using film as a form of social analysis'.[42] He met Carter a few days later:

> A long and stimulating talk with E.J. Carter, librarian at the R.I.B.A. who seems to be mixed up in a lot of interesting things. Explained setting up of the 1940 Council. Nobody on it with knowledge of public relations. He is to try for the setting up of a Sub-Committee with full powers to co-opt outside + explore the film position. The Council . . . will have two aims: research material now for future planning: publicity to arouse consciousness among ordinary people of their needs for proper social environment. Films to be used in latter campaign.[43]

We should note Rotha's language here: he is concerned with the needs of ordinary people for 'proper social environment', that is, one planned by appropriate experts, but at this stage there is no explicit reference to scientists. Carter became a crucial contact for Rotha in the years that followed, and it was Carter who brought him into contact with a group of radical scientists, the Tots and Quots.

---

39 E.J. Carter, 'The Limits of the 1940 Council's Mandate', covered by note to Sir Ian MacAlister, 30 August 1940, CaE/2/2, E.J. Carter Papers, RIBA Library Manuscripts and Archive Collections, London.

40 Published in illustrated book form as R. Tubbs, *Living in Cities* (Harmondsworth, 1942). See also Anon., 'Living in Cities', *Architectural Review*, June 1941, 143–4. Rotha diary, 7 May 1941.

41 Stevenson, 'Planners' Moon' (n. 3), 69–70.

42 Rotha diary, 2 and 3 October 1940.

43 Ibid., 4 October 1940.

The 'Tots and Quots', named by Haldane and Hogben, had originally met between 1931 and 1933. They had gathered over dinner more or less monthly, with discussions led by invited speakers about the economic and social relations of science. Members had included the scientists J.D. Bernal, Haldane, Hogben, Huxley, G.P. ('Gip') Wells, Solly Zuckerman and the economists M.M. Postan, Roy Harrod and Hugh Gaitskell.[44] Moves to re-establish the group had commenced around the time of the Munich crisis 'to alert scientists to the dangers that lay ahead'. Zuckerman, Bernal and Huxley had prepared a memorandum, published in *Nature*, urging the full involvement of scientists in the anticipated war.[45] The first reconvened Tots and Quots meeting occurred on 23 November 1939. Their most visible product was the July 1940 Penguin Special *Science in War*, published anonymously, but co-authored by 25 scientists and allies, including Zuckerman, Bernal, Crowther, Huxley, C.D. Darlington (then Director of the John Innes Horticultural Institute) and Carter, who attended from December 1939.[46] Publisher Allen Lane, present at the dinner of 12 June 1940, had 'accepted the challenge' to publish a text if it could be produced within a fortnight.[47]

Just a few days after they first met, Carter arranged for Rotha to see Zuckerman and Darlington, known to Rotha as contributors to *Science in War*. It is worth quoting Rotha's diary entry from his first meeting at some length:

> Got down to discussing films as a medium for publicising scientific progress. . . . Z very alive to possibilities; Carter anxious to formulate plans. As we talked, it became clear to me that there is an immensely important opportunity here of getting some of these specialists, each brilliant in his own field, into a group and setting them about the job of planning for the World Beyond War.[48]

Rotha was evidently most stimulated by the conversation. Although he had known several members of the Tots and Quots since the mid-1930s, this meeting with Zuckerman marked the point at which the social role of science entered into how he thought and wrote about the social role of documentary. He continued:

> If we take all the fields most closely affecting social environment – nutrition, medicine, agriculture, economics, social science, architecture, etc – + picked-out

---

44 S. Zuckerman, *From Apes to Warlords: An Autobiography, 1904–46* (London, 1988), 60–61 and 393–5.

45 Ibid., 101–2; and see unsigned editorial, 'Science and National Service', *Nature* 142 (1938), 685–7.

46 Tots and Quots papers, SZ/TQ/1/3/, Zuckerman Archive, University of East Anglia (hereafter 'Zuckerman Archive'). A full list is also given in Zuckerman, *From Apes to Warlords* (n. 44), 111–12.

47 Anon., *Science in War* (Harmondsworth, 1940).

48 Rotha diary, 8 October 1940.

> the most progressive worker in each, added some of those who are working in touch with 'public opinion' we could form a group of people who could formulate a plan for better living. That plan could be publicised – by press, film and radio, with the intention of creating the need for its being carried out when the war is suspended.[49]

He spelled out the role of publicity in the revolutionary implications of such a scheme under the special circumstances of war:

> If we can create an awareness in the minds of people of how much better they could live their lives if the full resources of science, industry, agriculture + architecture, etc [were made] available and made use of, nothing would stop those people getting them. But the resources have got to be known, not only the resources but their applications + publicity is necessary for this vast job. It can be done – + now with the stimulus of war to urge the necessity of a progressive peace, it must be done.[50]

This will to create public awareness of the world that could be created by science became highly significant for the subjects Rotha chose to represent in many of his films after this date.

Rotha films: social aspects of the war, post-war problems and planning

The changes in staff at the Ministry of Information mentioned earlier culminated for the documentarists with the appointment of one of their number, Arthur Elton, as Film Officer in January 1941.[51] At the very end of 1940, Rotha was able to set up a company and start making films again.[52] His company's first projects included films on salvage, blood transfusion, school education and diphtheria immunisation. The possibilities opened up by his new associates were uppermost in his mind as he recorded in his diary in January,

> The MOI films – those that we choose to accept – will be a sheet-anchor – but I want to explore the post-War field of reconstruction. There is the Science in War group plan + the '1940' Committee to pursue: and this interests me a great deal. Here is a real social growing point for documentary.[53]

The press release announcing the company stressed that 'documentary film is a medium through which audiences both at home and overseas may be shown

---

49 Ibid.

50 Ibid.

51 See Swann, *British Documentary Film Movement* (n. 4), 150–60; E. Sussex, *The Rise and Fall of British Documentary* (Berkeley, 1975), 121.

52 Rotha diary, 22 December 1940 and 8 January 1941.

53 Ibid., 8 January 1941.

how the Britain-to-come may be discussed and planned even while a war is being fought', and that there would be cooperation with experts:

> The Unit will establish a policy of producing films about progress in the fields of education, health, medicine, housing and the social sciences in terms of both war and post-war problems. Its research will be directed to this end, being based on work conducted over ten years in co-operation with specialists in these fields.[54]

Rotha published an advertisement in *Documentary News Letter*, which carried a quotation from Julian Huxley's *On Living in a Revolution*: 'The most vital task of the present age is to formulate a social basis for civilisation, to dethrone economic ideals and replace them by human ones'.[55] Rotha was not alone in believing that science films would benefit wartime propaganda; at the same time as the start-up of Rotha's company, Sidney Bernstein from the Ministry of Information had requested from Huxley a list of scientific subjects for films. Huxley's response included two subjects about which Rotha later made films: blood transfusion and diet in wartime.[56]

At the end of February 1941, Rotha attended a Tots and Quots meeting where he voiced his sense of the need for the Ministry of Information to move towards films concerned with post-war reconstruction and the need for an advisory body to his company to achieve this.[57] He recorded: 'The obvious use of the documentary film for interpreting scientific work was explained and, I feel, useful contacts were made'.[58] In early March, the 1940 Council formed a subcommittee to deal with films about reconstruction; its members included Huxley, David Owen (a leading figure in PEP), Carter, Ward, Calder, L.V. Easterbrook (agriculture correspondent of the *News Chronicle*) and the documentarist Basil Wright.[59] One of Rotha's first actions was to show this group *The Face of Britain*.[60]

---

54 P. Rotha, 'Documentary films yesterday and today', press release appended to P. Rotha to J. Huxley, 11 February 1941, Huxley Papers, Rice University, Houston, Texas (hereafter 'Huxley Papers').

55 *Documentary News Letter*, March 1941.

56 The full shortlist considered for production was: 'blood transfusion, diet in wartime, food preservation, new sources of energy, hormones, training of airmen, plastics, industrial psychology, genetics of race, new methods of wound treating, textile research and meteorology'. S. Bernstein to J. Huxley, 7 February 1941, SZ/TQ/251, Zuckerman Archive. See also Huxley to Rotha, 13 February 1941, Huxley Papers.

57 Tots and Quots papers, SZ/TQ/1/7, Zuckerman Archive.

58 Rotha diary, 2 and 3 April 1941.

59 Ibid., 11 March 1941.

60 Ibid., 4 April 1941.

Rotha's contacts with Calder continued throughout this period. For example, his diary for 17 February 1941 records 'broken off here because Ritchie Calder comes along with his two new pamphlets "The Lesson of London" + "Start Planning Britain Now"'. The latter pamphlet was written to show that the destruction wrought by bombing could provide the opportunity for improved living conditions. But Calder was keen to make the broader connection: 'physical planning must not be separated from economic and social planning or from the cultural and the ethical'.[61] Calder saw the London of the Blitz as the launch pad of substantial social change. His *The Lesson of London* is more vividly rhetorical:

> We are not fighting for 'Democracy' ... with a capital 'D' ... not for the 'Democracy' which was once a coin brightly minted from the gold of men's aspiration, and to which the French Revolution gave its superscription 'Liberty, Equality and Fraternity', but which has been so long in the pockets of the Ruling Class that its device has worn off ... not for a Democracy of privilege and slums. No, the people who are holding the Front Line are fighting and suffering for a new democracy, which they can understand, and to which they are giving the meaning 'democracy' without the capital. The coin is being reminted, its true worth fresh[ly] assayed in the fires of human endurance, and its superscription is that simple remark of the docker: 'We are all in it together'.[62]

In a similar way, Rotha's left politics expressed themselves in a particular interpretation of reconstruction that included a strong sense of social revolution. As he noted in his diary,

> The big industrialists and manufacturers will only go so far in the war effort so as to secure a so-called 'victory' – but they still care [most] for profits.... The money makers will ... prefer a negotiated peace, in which they hope to retain their money and their status, rather than a peace in which a more equalised method of living will be demanded.[63]

The contacts with the 1940 Committee and the Tots and Quots had their most explicit result in a film Rotha planned with the title *Science and War*, first mentioned in his diary on 7 May 1941, for which first draft scripts were being composed the following month.[64] Writing to his friend, the novelist Eric Knight, in July, he sketched the socially useful complexion he put on science:

> There's a biggish [film] on Science and War and just how science has been given its chance in Britain since September, 1939. And by science I don't mean guys

61 P.R. Calder, *Start Planning Britain Now* (London, 1941), 20.

62 P.R. Calder, *The Lesson of London* (London, 1941), 124.

63 Rotha diary, 19 June 1941.

64 Ibid., 22 June 1941.

> splitting atoms or even getting a fighter to fly at 400 instead of 399 mph. I mean the fact that the population is for the first time in history getting an adequate diet, that agriculture is being approached scientifically for the first time, that free milk is being given to millions of kids, that inspection of bomb damage is revealing facts about building construction that will vitally affect reconstruction, that from now on science must be an integral part of society.[65]

He intended that the film should end with sequences from the BAAS Science and World Order conference, largely organised by Calder. It took six paragraphs for Rotha to outline the film's vision of science. In abbreviated form, what he said was:

> [W]e start with a survey of world needs in raw materials and food and [we] interpret war aims. . . . The conflict of ideologies goes hand in hand. Then we naturally develop the theme of science and society. We take science in three countries in the post-[First World War] years; in Britain, and USSR and in Nazi Germany. . . . So to the War itself. Forget the offensive war of attack with bomb and warplane. What matters to our film is that it has taken a war to make the British Government realise that their attitude to food, agriculture, fuel supply, medicine, etc., must be scientific if we are to win this war. . . . In wartime, scientists go out and discover for themselves what is happening. . . . Out among the ruins and fields they go; see for themselves the effects of bomb-blast on human beings, see for themselves what it means to reclaim 1000 acres of marsh-land. This is important and it has got to stay. The technologist, the scientist, the architect, the specialist has at last been called in by the British state. Those are the men who are going to plan for a new Britain. . . . [I]f this film can educate the great mass of people just what resources of science lay ready for the common good, perhaps the great mass will see that they are made use of.[66]

Serious work was done on the film, and by early August 1941 W.G. Bennett and Michael Orrom, working for Rotha's company, had completed a first script. Rotha sent it out for discussion by a wide group of people from the groups he had been mixing with in the first eighteen months of the war, including Bernal, Carter, Crowther, Huxley, Hogben and Zuckerman.[67]

Science and War never went beyond the first script stage, for reasons that are difficult precisely to establish. It cannot have helped that Orrom, who had

---

65 Rotha to Knight, 19 July 1941, 3, Box 26, Rotha's letters to Eric Knight, Rotha Archive.

66 Rotha to Knight, 19 July 1941, 3, Box 26, Rotha's letters to Eric Knight, Rotha Archive.

67 Rotha to Carter, 4 August 1941, Ca E/2/3, E.J. Carter Papers, RIBA Library Manuscripts and Archive Collections, London. Rotha to Zuckerman, 7 August 1941, SZ/TQ/2/3/3, Zuckerman Archive; Zuckerman to Rotha, 12 August 1941, SZ/TQ/2/3/4, Zuckerman Archive. There are copies of the script in both archives.

been developing the script, was conscripted to the radar corps in August 1941,[68] nor that Calder was involved with the joint burden of organising the Science and World Order conference and starting work at the Foreign Office's Political Warfare Executive from August.[69] Ultimately, it seems that the Ministry of Information cancelled the film: Rotha's speech to the first annual meeting of his film company commented that 'we depended largely on contracts from the Ministry of Information, which body was still in a state of sorting itself out. Much time was wasted on films allocated to us from this body which were later cancelled'.[70] Presumably *Science and War* was one such. Michael Orrom's later recollection was that 'the project had to be shelved because the subject matter was so heavily restricted by wartime security conditions'.[71] In August, Rotha had suggested to the Ministry, '[T]hey allow me to make a short [film of the Science and World Order conference] for immediate priority release as well as sequence for "Science and War" film. If so, we will use Ritchie Calder as narrator and compere to explain Meeting'.[72] In the end, *They Met in London* was released early in 1942. It was a documentarist's version of a newsreel, shot by the newsreel company British Paramount News, and only hinted at the scope and ambition of *Science and War*.

Rotha and Neurath

In parallel with the development of *Science and War*, Rotha met, and began working with, Otto and Marie Neurath, who had been released from internment in February 1941. They had settled in Oxford under the protection of the Marxist economist and political theorist G.D.H. Cole, with the intention of restarting their Isotype work. Rotha wrote via Neurath's solicitor, saying that Calder had alerted him to Neurath's presence in Britain, regarding possible work on the blood transfusion film: 'having admired your work in pictorial symbolism for many years, I feel that there is no-one more qualified to help in this matter than yourself'.[73] Neurath responded, '[T]he instance of blood transfusion offers many possibilities for applying our Isotype method to science animated films.

68 Huxley to Rotha, 6 and 16 August 1941, Huxley Papers.

69 Rotha diary, 17/18 August 1941; *The Times*, 2 February 1982, 12.

70 P. Rotha, 'Chairman's report to shareholders', 3 May 1942, folder 2–3, Box 72, Rotha Archive.

71 K. McGahan, 'Conflict and Confluence: Michael Orrom', in P. Russell and J.P. Taylor (eds), *Shadows of Progress: Documentary Films in Post-War Britain* (London, 2010), 176.

72 Rotha diary, 17 August 1941.

73 Rotha to Neurath, 7 April 1941, Isotype Collection, Manuscript collection, University of Reading.

Our International Foundation for Visual Education was always interested in problems of health education and made experiments years ago in animated films'.[74] Rotha paid a visit to Oxford on 12 May 1941.[75]

The auguries were good for the meeting, given Rotha's strengthened commitment to the social role of science and planning and Neurath's long-standing commitment to the same principles. In fact, the first meetings were marked by a mutual awe. Marie Neurath recalled,

> Rotha came to see us in Oxford. But we found it impossible to receive him in our combination bed sitting room with its little table in the middle.... So we asked for a room in GDH Cole's library and quickly made it look as if we had used it for some time.[76]

Rotha described the arrival of the Neuraths as 'exciting'. He wrote in his diary:

> On Tuesday I went to Oxford to see the Neuraths in their small converted room, probably once a housemaid's bedroom, now occupied by this good-humoured, large Viennese professor and his charming blue-eyed wife. We talked all day of methods of using Isotype for the backbone of the Blood transfusion film and for the salvage film.... [They] are so intelligent, so cultured. His invention of 'picture language' or Isotype is perhaps one of the most revolutionary and important steps in interpretive education.... He has so many ideas and is so full of initiative and energy.[77]

Rotha continued to be struck by Neurath's long-standing concern with planning. More than a year later, he wrote to Knight:

> The great thing about Otto Neurath is that he is a brilliant economist first and a diagram king second. The more I talk with him, the more I am impressed with the man's brain. His ideas on planning are sane and inspiring. He has been at all of this for twenty years and, in his own field, is as brilliant as Orr and the others.[78]

As we shall see, Rotha, Neurath, Knight and Orr would soon be working together on one of Rotha's most significant films of the war, *World of Plenty*.

---

74 Neurath to Rotha, 9 April 1941, Ibid.

75 This date is calculated from Rotha diary, 16 May 1941.

76 Quoted in O. Neurath, *Empiricism and Sociology*, ed. M. Neurath and R.S. Cohen (Dordrecht, 1973), 72.

77 Rotha diary, 16 May 1941.

78 Rotha to Knight, 6 July 1942, 3, Box 26, Rotha's letters to Eric Knight, Rotha Archive. On Neurath's interest in planning generally, see George Reisch, 'Planning Science: Otto Neurath and the International Encyclopaedia of Unified Science', *British Journal for the History of Science* 27 (1994), 153–75.

Rotha and the Neuraths' Isotype Institute collaborated on 17 films during and after the war (collaboration continued between Rotha and Marie Neurath after Otto's death).[79] Some of these, including the five-minute salvage film *A Few Ounces a Day* (1942), were pure exercises in the use of Isotypes; others, such as *Blood Transfusion* (1941), were on explicitly scientific themes. And just a few expressed the themes intended for *Science and War*. One such was *World of Plenty*, the film Rotha made with Knight and Boyd Orr with some assistance from Calder.[80]

The film was suggested by Rotha to the Ministry of Information in October 1941 (very soon after *Science and War* must have been cancelled) and first shown in May 1943. As with *Science and War*, he turned to his wider circle of associates, including Huxley, Calder and the Labour politician and old contact Stafford Cripps, to assist in developing and finalising the script.[81] Boyd Orr not only featured centrally in the film as expert on nutrition science and visionary as regards food policy, he also entirely rewrote the last third of the film. Neurath had a major hand in developing the film, with Isotypes featuring strongly in the exposition of its argument.[82]

World of Plenty is structured as a three-part historical dialectic and tells the story of food production and distribution before, during and after the war. In a technique that Rotha adapted from the New York 'Living Newspaper' stage shows, the film uses a multi-voice commentary, in which the argument is carried on the soundtrack by several voices in discussion with each other.[83] The first part, 'Food – as it was', concentrates on pre-war problems in Britain and America, especially during the Depression. It is centred on the theme that pre-war food production was not adequate to provide for the health needs of populations and speaks of the injustice that some populations, even in the wealthiest nations, were ill-nourished. The second part, 'Food – as it is', deals with problems of shipping food in wartime, President Roosevelt's Lend-Lease scheme and food rationing policy in wartime Britain. The final part, 'Food – as it might be', stresses the

79 For a complete list, see J. Edwards and M. Twyman, *Graphic Communication through ISOTYPE* (Reading, 1975), 45.

80 Calder had admired Boyd Orr ever since their first meeting, an interview for Calder's book *The Conquest of Suffering* (London, 1934). (This book has an introduction by J.B.S. Haldane.) Accordingly, Calder gave publicity to Orr's survey of surveys, *Food, Health and Income* (London, 1936), which argued that half the British population could not afford a properly nutritious diet.

81 Rotha diary, 17 May 1942.

82 See T. Boon, 'Agreement and disagreement in the making of *World of Plenty*', in D.F. Smith (ed.), *Nutrition in Britain* (London, 1997), 166–89.

83 See D. Pearson, 'Speaking for the Common Man: Multi-Voice Commentary in *World of Plenty and Land of Promise*', in P. Marris (ed.), *Paul Rotha* (London, 1982), 64–85.

links between nutrition and health and the availability of scientific knowledge by which – it argues – devastated farmlands could be made productive again. It urges the necessity for a world food plan based on 'supplying human needs by making the resources of the whole world available to all the peoples of this world'. When one of the main voices bursts out 'But this is revolutionary!', Boyd Orr replies with great deliberation: 'Tell me, what are we fighting for if not for something revolutionary? What do people like you and me hope to get out of this war if not a better world?'[84]

Isotypes are used throughout the film to present factual information dramatically on behalf of the protagonists. For example, an Isotype world map features at the beginning of the film; first it shows the major producing nations for different foodstuffs, quickly followed by a variant showing where the world population was concentrated (Figure 8.2); finally, trade in a few key commodities between net producers and net consumers is shown by animating the map. This is undertaken with the help of some verbal interplay between the two main voices, 'Newsreel', who is presented as wanting to do a straightforward film on food supply in wartime, and 'Man in the Street', a sceptical, questioning, everyman who was intended to represent the audience's voice in the film, holding other speakers to account. Among Isotypes later in the film are examples showing income distribution in the USA and Britain; incidence of rickets and tuberculosis; and how Boyd Orr's World Food Plan was intended to work.

In *Land of Promise*, the other major collaboration between Rotha and Otto Neurath, there was perhaps even more of a meeting of minds between the two men, especially in terms of their thinking about planning. By mid-1942, Rotha's views had become more critical as regards 'planning from above', as he summed up in a letter to Knight in terms echoing Carter's aims for the 1940 Council:

> So many of the would-be planners are people who think only in terms of big-scale bricks-and-mortar planning. Rebuild our cities and roads, they scream, and disease and poverty will vanish. This represents a most dangerous slogan. I have a particular slogan at the moment – 'Planning for Public Health' – which is the theme behind these medical pictures. These countries cannot have some aesthetic-architectural plan superimposed on them from above without proper research into what human needs require. Orr is very sound on this in his 'wee bookie'.[85]

As is clear from this reference to Orr's *Fighting for What?*, Rotha's sense of the proper form of planning for reconstruction continued to be nurtured by the developing views of others in his circle. Rotha asserted in his diary that

---

84 After the war, Orr was appointed Director General of the UN Food and Agricultural Organization (FAO). He resigned in 1948 after his World Food Plan had been rejected.

85 Rotha to Knight, 6 July 1942, 5, Box 26, Rotha's letters to Eric Knight, Rotha Archive.

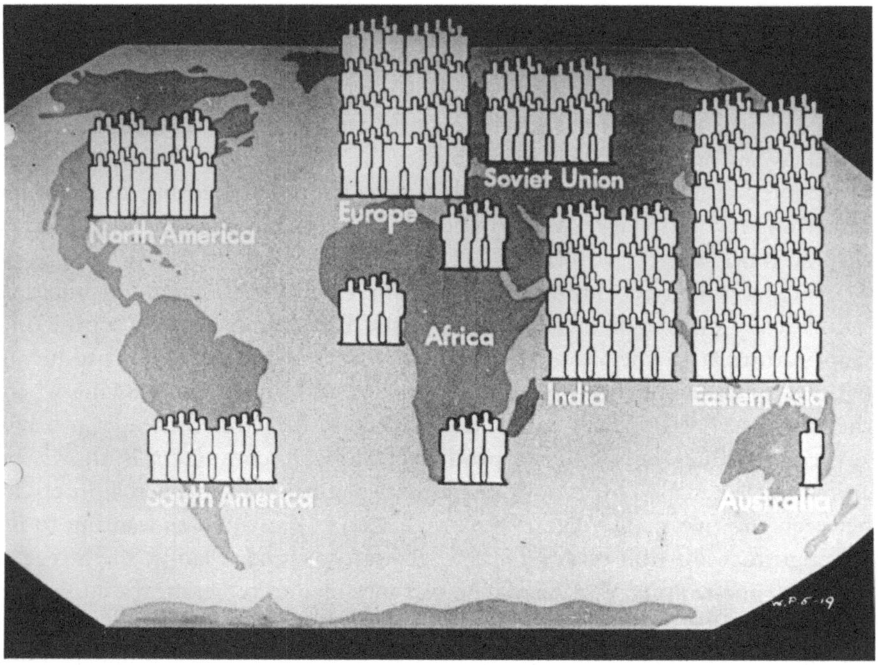

Figure 8.2 Distribution of the world's population: Isotype map.
Source: From *World of Plenty* (Crown Copyright, supplied by Imperial War Museum).

Orr's book – which he had in manuscript – 'has meant much to me'.[86] Orr argued that 'the rise of so many groups of planners having no connection with government or any of the great political parties is an indication of the widespread desire for dramatic changes in the economic system. The people of Great Britain will no longer tolerate unemployment, poverty and slums'. Orr's was a type of technocratic progressivism: '[W]hen the enemy is defeated, [the common people] must fight and work with the same tenacity to build the new world. They will have fighting and working by their side men of vision and goodwill in all classes. This is the people's war'. Yet for Orr, in the end it is the scientists who will enable the new world to be created: 'Men of science have now given us the physical means of attaining the new world order which our poets and prophets have seen only in their visions'.[87]

86 Rotha diary, 13 July and 10 May 1942.

87 J. Orr, *Fighting for What? To Billy Boy and All the Other Boys Killed in the War* (London, 1942), x, xiv and 11.

Neurath, on his part, had long-standing experience in central planning, which he had applied in the Bavarian revolution and in interwar 'Red Vienna'.[88] He became increasingly committed to actively engaging citizens in decision making about planning. His use of Isotypes as a means of visual communication not only had the purpose of making technical knowledge more accessible, but also of empowering lay people to apply that knowledge in their own lives. By thus enabling people to make their own rational choices, the decision making process could, he argued, be broadened from small circles of experts to a well-informed general public.[89]

In August 1941, in the midst of developing ideas for *Science and War*, Rotha had begun thinking about a film on a broad canvas, touching on planning in general. He had already been in contact with Reith, the Minister of Works, who asked him to write a memorandum on the potential use of films for reconstruction propaganda.[90] There was mention of a collaboration with Thomas Sharp, the author of the 1940 Pelican *Town Planning* then working for Reith, who had already submitted a script to the 1940 Council film committee.[91] In May 1942 there was talk of Rotha being commissioned by the gas industry to do a planning film, the idea having been sold to them largely on Sharp's script. Rotha commented: 'If I do it, Sharp's script must gradually be discarded + a real planning script substituted. Planning for human needs – *that's* all we can plan for'.[92] This film was finally commissioned in October 1943, and emerged after a long gestation – and with the collaboration of Neurath – as *Land of Promise* in 1946 (Figure 8.3).[93]

Land of Promise was in many ways the final expression of Rotha's long-nurtured wish to make a film on themes of urban planning, architecture and housing. It is, in a way, the *World of Plenty* of housing. Like the earlier film, it uses a historical dialectic of pre-war, wartime and post-war. It aims to show how the heightened level of planning needed in wartime could be extended into the post-war world to overcome the problems of slum housing that had been a subject for documentary films since Arthur Elton and Edgar Anstey's *Housing Problems* in 1935. Once again, it uses Rotha's multi-voice commentary technique; it features John Mills, as 'The Voice'– an everyman figure speaking

88 Neurath, *Empiricism and Sociology* (n. 76), 18–28.

89 S. Nikolow, 'Planning, Democratization and Popularization with ISOTYPE, ca. 1945: A Study of Otto Neurath's Pictorial Statistics with the Example of Bilston, England', in F. Stadler (ed.), *Induction and Deduction in the Sciences* (Dordrecht, 2004), 299–329.

90 Rotha diary, 7 May 1941.

91 Ibid., 18 August 1941.

92 Ibid., 13 July 1942.

93 Ibid., 18 October 1943.

Figure 8.3 Woodcut illustration, probably by Peter Strausfeld.

Source: From Academy Cinema leaflet for showing of *Land of Promise* (Isotype Archive, University of Reading).

on behalf of the audience, Miles Malleson as 'Mr Know-it-all', Marjorie Rhodes as 'The Housewife' and Herbert Lomas as 'History'. In a departure from *World of Plenty*, however, Isotype is here not merely used as a visual technique to convey statistical information, but as a character in the film. Using the logotype of the figure holding a sign, 'Isotype', voiced in slightly robotic tones, is introduced, along with the other players, at the start of the film:

> I am ISOTYPE. I use symbols to make diagrams. Here is my symbol for a family: father, mother and children. Here are houses people live in. Different kinds of houses. We shall be needing builders because we shall be building houses. I am here to help. I am ISOTYPE.

Throughout the film, animated Isotype symbols are used to convey some of the more abstruse arguments, accompanied by percussive music. Because they are applied using the consistent rules characteristic of the system, they serve to anchor and clarify the very dense web of the film's subject matter. Whereas 'History' draws on official evidence to convey where housing problems had come from, 'Isotype' is a voice of factual information, proposing for example that, if aircraft factories are organisations for turning raw materials into fighters and bombers, then they can also be used to turn similar raw materials into prefabricated houses in the post-war period.

Land of Promise is thematically very much the culmination of Rotha's and Otto Neurath's collaboration. With the personification of Isotype, it marks the most significant use of the technique in documentary films. It marked the convergence of the two men's commitment to social planning in its application to the design and development of housing and towns. In this sense it drew on Neurath's connections with the architects and artists of the *Bauhaus* as well as his experience in the Museum of Economy and Society in Vienna, and on Rotha's contacts with the architectural profession dating back a decade.[94] Along with *World of Plenty*, it is a manifestation of Rotha's argument for the use of documentary as a tool of citizenship, as he argued at the BAAS conference on Science and the Citizen in 1943:

> Not only should films about science give facts, but they should present facts in such a way that may well invoke a call to action. . . . If [they] are going to have real social purpose, they must give people facts and information not drily but dramatically presented. Above all, they must create a desire among the people for science to be used for *all the people*, not for the privileged few alone.[95]

Neurath expressed similar views in his last, unfinished paper, 'Visual Education: Humanisation Versus Popularisation'. Rejecting older attitudes towards popularisation as too paternalistic for a modern citizenry, he instead stressed 'humanisation'. This meant providing people with the knowledge they needed to make their own rational choices and become active participants in democratic

---

94 On Neurath's links with the *Bauhaus*, see especially P. Galison, 'Aufbau/Bauhaus: Logical Positivism and Architectural Modernism', *Critical Inquiry* 16 (1990), 709–52.

95 P. Rotha, 'Interpreting Science by Film', *Advancement of Science* 2 (1943), 306, Rotha's emphasis.

decision-making about planning. Only by bringing together planning and public education, he argued, could planning become a truly democratic process.[96]

Conclusion: visuality and modernity

Leo Charney and Vanessa Schwartz assert in their collection *Cinema and the Invention of Modern Life* that, of all emblems of modernity, 'none has both epitomized and transcended the period of its initial emergence more successfully than the cinema'.[97] We may add that an emphasis on the visual was often characteristic of the type of modernism that saw science as providing the means to rationalise society.

The emphasis on the visual, as a core constituent of the modernist project, is seen in both Rotha's and Neurath's work. In the case of Neurath, we need only consider that Isotype is a universal language, certainly, but a *picture* language, not a verbal language such as Esperanto, Basic English or Hogben's 'Interglossa'. With Rotha, the emphasis on the visual had its origins in his commitment to the Russian dialectical montage of Eisenstein and Pudovkin. Rotha's application of the technique, in turn, received the approval of Pudovkin, who wrote to Rotha after he saw *World of Plenty* in 1945, praising the film and talking about it in terms of montage theory:

> [N]o other art than cinematography has the opportunity of so forcefully showing the ties linking facts. The cinema can make the abstract relations between facts not only audible (as in speech) but also visible. I always remember the usual expression employed by some one who wishes to drive home an argument.... 'I have seen it with my own eyes!'[98]

There is a similarity between the fusion of technique and conveying of meaning in montage theory and Neurath's assertion:

> The ISOTYPE method ... shows connections between facts instead of discussing them. Impressive visual aids do not merely act as illustrations or as eye bait ... they are parts of the explanations themselves.[99]

Neurath's and Rotha's shared stress on visual communication, expressed so powerfully in films such as *World of Plenty* and *Land of Promise*, was joined

---

[96] O. Neurath, 'Visual Education: Humanisation Versus Popularisation', in E. Nemeth and F. Stadler (eds), *Encyclopedia and Utopia* (Dordrecht, 1996), 245–335.

[97] L. Charney and V. Schwartz (eds), *Cinema and the Invention of Modern Life* (Berkeley, 1995), 1.

[98] V. Pudovkin to P. Rotha, 19 April 1946, folder 5, Box 66, Rotha Archive.

[99] O. Neurath, *Modern Man in the Making* (London, 1939), 7–8.

by a common commitment to use it to communicate to the general public the transformative power of science and planning. Their shared project found a particularly compelling expression in their cinematic collaborations. But, as we have seen, these films were the culmination and fusion of concerns they had separately developed, working in different social circles and contexts, over two decades.

Acknowledgements

I owe a debt to the organisers and audiences of several conferences and seminars who provided me with opportunities to explore aspects of this material. I should like to thank the archivists at the Rotha Archive (University of Los Angeles), the Royal Institute of British Architects, Zuckerman Archive (University of East Anglia), Huxley Papers (Rice University), National Library of Scotland and the Isotype Archive (University of Reading) for permission to consult and quote from collections in their care. I acknowledge with pleasure permission granted by the executor of the Rotha estate to reproduce items in the Rotha Archive. I owe a debt to my colleagues at the Science Museum and in the Science Museum Library. For Figure 8.1, I thank BFI Films, Stills and Designs and Viewtech Film and Video; for Figure 8.2, I thank Imperial War Museum and Film Images; and for Figure 8.3, I thank the Isotype Archive, University of Reading, and Peggy Strausfeld.

Chapter 9

Unifying science and human culture

The promotion of the history of science by George Sarton and Frans Verdoorn

Bert Theunissen

In the late 1970s I attended a seminar at what was then called the Utrecht Biohistorical Institute. As a biology student majoring in ecology and used to fieldwork in constantly bad weather, I was struck by the humanistic serenity of the institute's atmosphere. It was housed in an ancient canal-side building shaded by elm trees in the old town, with parlour-like rooms and creaking wooden floors, ceiling-high bookcases and portraits of famous biologists decorating the walls. Career decisions can be made on the basis of small details: I knew instantly that I had found my destination and switched from ecology to history of biology.

Little did I know then that the distinctive character of the institute was all that was left of the so-called biohistorical ideology from which it still derived its name. The institute had, only a few years earlier, changed its programme from biohistory to history of biology, and small though this change of name may seem to be, it actually implied a major turnabout in its objectives.

Biohistory had been the invention of Frans Verdoorn (1906–1984), botanist, entrepreneur, publisher, bibliophile and self-taught historian.[1] In 1940 he had emigrated to the United States with his publishing company, but in the mid-1950s Utrecht University invited him to return to the Netherlands. Verdoorn accepted the newly created position in biohistory and the institute that was part of the deal. He sold his company and repatriated in 1958, bringing with him a superb biohistorical book collection and detailed plans for a biohistorical research and documentation centre.

---

[1] For biographical information, see A.P.M Sanders and H. de Vries, 'Frans Verdoorn, Plant Scientist and Biohistorian (with a Selective List of his Publications); a Concise Chronology', in P. Smit and R.J.Ch.V. ter Laage (eds), *Essays in Biohistory and Other Contributions, Presented to Frans Verdoorn on the Occasion of his 60th Birthday* (Utrecht, 1970), i–xvi; P. Smit, 'Van Biografie tot Biohistorie', *U.B.I. Wendingen. Incidentele Mededelingen van het Biohistorisch Instituut te Utrecht*, N.S. 4, October 1976; A.P.M. Sanders, 'In Memoriam: Frans Verdoorn, Bryologist, Publisher and Biohistorian', *Janus* 71 (1984), 165–79; S.P. Gradstein and P.W. Richards, 'Obituary Verdoorn', *Journal of Bryology* 14 (1986), 203–13; J. Ewan, 'Eloge: Frans Verdoorn, 24 July 1906 – 18 May 1984', *Isis* 78 (1987), 415–16.

These plans had been long in the making and had slowly matured during the years that Verdoorn ran his publishing company in Waltham, Massachusetts, near Boston. His principal source of inspiration in their development had been the writings of George Sarton (1886–1956), the editor of *Isis*, who lived in nearby Cambridge and worked in Harvard's Widener Library. The two men knew each other well. Verdoorn's company published Sarton's guide to the history of science, *Horus*,[2] and they were both members of a club of which Verdoorn was one of the founders, the Boston Biohistorical Club. Verdoorn considered Sarton's guidance as having been essential for the development of his biohistorical ideas: '[M]y whole idea of viewing the borderlands between the biological and humane sciences [i.e. biohistory] as a whole', he once wrote to Sarton, 'is mainly the result of your teaching and writing'.[3]

Although Verdoorn would never become a true Sartonian, the two men did have more in common than their interest in the history of science. They were pioneers of the discipline and contributed, however differently, to its professionalization. Ultimately, however, they were both unsuccessful in achieving the more specific aims they had set themselves. Sarton's pleas for a New Humanism failed to make a lasting impression and his unfinished magnum opus, the *Introduction to the History of Science*, is only rarely consulted by modern historians.[4] Until recently, his other works were cited predominantly in historiographical studies the tone of which was often critical.[5] Verdoorn's scant

---

2 George Sarton, *Horus: A Guide to the History of Science* (Waltham, MA, 1952).

3 Verdoorn to Sarton, 13 March 1956, Verdoorn Papers (hereafter VP), Institute for History and Foundations of Science, Utrecht University.

4 George Sarton, *Introduction to the History of Science*, 3 vols in 5 parts (Baltimore, 1927–1948).

5 See for instance Arnold Thackray and Robert K. Merton, 'On Discipline Building: The Paradoxes of George Sarton', *Isis* 63 (1972), 473–95; Tore Frängsmyr, 'Science or History: George Sarton and the Positivist Tradition in the History of Science', *Lychnos, Annual of the Swedish History of Science Society* 1973–1974 (Uppsala, 1975), 104–44; Nathan Reingold, 'History of Science Today, 1: Uniformity as Hidden Diversity: History of Science in the United States, 1920–1940', *British Journal of the History of Science* 19 (1986), 243–62, esp. 249–52; Lewis Pyenson, 'What is the Good of History of Science?', *History of Science* 27 (1989), 353–89; and Lewis Pyenson, 'Inventory as a Route to Understanding: Sarton, Neugebaur, and Sources', *History of Science* 33 (1995), 253–82. See also Thomas Kuhn, *The Essential Tension* (Chicago, 1977), 148. Recent work is more contextualising and sympathetic: Lewis Pyenson and Christophe Verbruggen, 'Ego and the International: The Modernist Circle of George Sarton', *Isis* 100 (2009), 60–78; idem, 'Elements of the Modernist Creed in Henri Pirenne and George Sarton', *History of Science* 49 (2011), 377–94; Christophe Verbruggen, 'Het Egonetwerk van Reiner Leven en George Sarton als Toegang tot Transnationaal Intellectueel Engagement', *Belgisch Tijdschrift voor Nieuwste Geschiedenis* 28 (2008), 87–129; and Geert Somsen, 'Wetenschap als Werelderfgoed: George Sartons Internationalistische Geschiedschrijving',

historical studies have already fallen into complete oblivion. His biohistorical ideology did not gain a foothold either. The Biohistorical Institute changed its programme from biohistory to history of biology and, in the late 1980s, was merged with other disciplinary groups to constitute the Utrecht Institute for the History of Science.[6]

One of the reasons that historiographical studies have adduced for the limited success of Sarton's ideas is that they were already out of date during his lifetime. His New Humanism is seen as a personal creed that met with little response and played no important role in the dissemination of his ideas. Moreover, it seemed naive in a world that had witnessed the atrocities made possible by science during the First World War. Likewise, his progressivist views on science and its history, rooted in Enlightenment rationalism and Comtean positivism, must in the eyes of his contemporaries have seemed to belong to the nineteenth century. Finally, Sarton's historical work, characterised by a bibliographic and encyclopaedic approach that had been common among the nineteenth-century pioneers of the discipline, had lost its appeal for most twentieth-century historians. Sarton provided the profession with useful tools, such as a journal, introductory texts and bibliographic guides, but he failed to give the discipline a cognitive identity, a set of problems that guided its practitioners' research.[7]

A similar explanation can be given for the fiasco of Verdoorn's biohistorical ambitions. His biohistorical ideology harked back to traditional ideas such as the unity of the arts and sciences, and even though Verdoorn successfully prepared the ground for the institutionalisation of his new field, its content remained too vague to form the backbone of a feasible biohistorical programme.[8]

While such critical appraisals provide plausible explanations of why Sarton and Verdoorn's ambitious schemes did not survive, they are rather uncharitable.

in Frank Huisman, Nico Randeraad and Georgi Verbeeck (eds), *Geschiedenis is Overal* (Amsterdam, 2013), 104–121.

6 For an overview, see H.A.M. Snelders, 'History of Science Today, 2: History of Science in the Netherlands', *British Journal of the History of Science* 20 (1987), 343–8.

7 See note 5. Of the authors mentioned there, Pyenson's stance towards Sarton's work is the most sympathetic where his encyclopaedic and synthetic approach is concerned. On the other hand, by labelling him as a chauvinist, Pyenson fails to do justice to how Sarton, in a quite typical way for the period, combined nationalism and internationalism. See Pyenson, 'What is the Good' (n. 5), 376–8. For more on this issue, see below.

8 Bert Theunissen, 'Frans Verdoorn: op Sartons vleugels, in Linnaeus' voetspoor', *Tijdschrift voor de Geschiedenis der Geneeskunde, Natuurwetenschappen, Wiskunde en Techniek* 12 (1989), 149–66; Bert Theunissen and Rob P.W. Visser, 'History of Biology in the Netherlands: a Historical Sketch', *Tractrix: Yearbook for the History of Science, Medicine, Technology and Mathematics* 2 (1990), 141–57.

Not only are both men's ideas found wanting by measuring them against modern standards, it is also suggested, implicitly or explicitly, that these ideas were mere relics of a former age even during their lifetimes. Thus both men appear as isolated eccentrics who never quite managed to tune in to the spirit of their times.[9]

Now eccentrics they definitely were, but they were no living fossils. A more balanced view of their hopes and ambitions is obtained by taking their background as émigrés from two small European countries into account. There is a direct relation between this background and their conception of the history of science, and on this perspective their ideas were less idiosyncratic and outlandish than one might be inclined to think. It is also this relation that directly connects Sarton and Verdoorn to the theme of this volume: the unity of science. For it was the idea of an overall unity – of nature, of science and of humankind – conceptualised against a background of long-term political developments in their home countries that stimulated, perhaps even generated, their historical interest. More particularly, their careers point to the importance of scientific internationalism as one of the moving forces behind the development of the history of science in the twentieth century. Thus, while the appeal of their specific historical projects may have been limited and short-lived, Sarton and Verdoorn's careers illustrate important aspects of the history of the history of science.

History of science and world peace

George Sarton's mother died when he was only a few months old, and he grew up a lonely child.[10] His family background also played a role in this. The Sartons lived in Ghent in Flanders, a conservative, catholic, Dutch-language city.

---

[9] In a later article, however, Arnold Thackray showed that among the American patrons of the history of science in the first half of the twentieth century, Sarton's neo-humanistic convictions and internationalism (the European roots of which I shall explore below) were in fact widely shared: A. Thackray, 'The Pre-History of an Academic Discipline: The Study of the History of Science in the United States, 1891–1941', *Minerva* 18 (1980), 448–73.

[10] For biographical information, see May Sarton, *I Knew a Phoenix* (New York, 1959); and idem, *A World of Light* (New York, 1976); Lewis Pyenson, *The Passion of George Sarton: A Modern Marriage and Its Discipline* (Philadelphia, 2007); A.M. Simon-van der Meersch (ed.), *George Sarton's European Roots: Introduction to his Life and Times* (Ghent, 1984); Paul van Oye, *George Sarton: De Mens en zijn Werk uit Brieven aan Vrienden en Kennissen*, Verhandelingen van de Koninklijke Vlaamse Academie van Wetenschappen, Letteren en Schone Kunsten van België, Klasse der Wetenschappen, vol. 27, no. 82 (Brussel, 1965); Pyenson, 'Inventory as a Route to Understanding' (n. 5); Arnold Thackray and Robert K. Merton's entry on 'Sarton, George Alfred Léon' in *Dictionary of Scientific Biography*, vol. 12 (New York, 1975), 107–14 ; and the commemorative contributions to *Isis*, vol. 48, no. 3, 1957; vol. 75, no. 1, 1984; vol. 76 (1985), 470–99; and vol. 100 (2009), 58–107.

Figure 9.1 Signed portrait of George Sarton, aged around 60.
Source: Collection Institute for History and Philosophy of Science, Utrecht University. Reproduced with permission.

Sarton's father, however, a railroad engineer, spoke French, was a liberal and an anti-clerical freethinker. During his adolescence Sarton rebelled against the bourgeois milieu of his family. He experienced a *Sturm und Drang* phase, wanted to change the world, became a socialist and wrote poetry.

Sarton's isolation came to an end in the years that he studied chemistry and mathematics at the University of Ghent. He made some similarly idealistic friends, among whom his future wife, and together they dedicated themselves to the ideals of pure living, vegetarianism and socialism. The idea of unity, which would become the moving spirit of his life's work, made its first appearance, in a rather cryptic form, in his private journal from this period: 'I believe one can divide men into two principal categories: those who suffer the tormenting desire for unity and those who do not. Between these two kinds an abyss – the "unitary" is the troubled; the other is the peaceful.'[11]

[11] May Sarton, *I Knew a Phoenix* (n. 10), 40–41. On Sarton's student life, see Verbruggen, 'Egonetwerk' (n. 5).

In the years before the First World War, after having considered careers as a mathematician and a writer, Sarton settled for the life of a private scholar and began to devise plans for the publication of a synthetic history of science and an international history of science journal – the first issue of *Isis* was to appear in 1913. The aims he set himself testify to his high-strung expectations from this line of work. His history of science aimed at nothing less than to help putting an end to international political rivalries and to show the way toward global cooperation and peace. More concretely, Sarton hoped that his work might help in bringing about a new world that was freed from dilemmas such as that in which Belgian politics found itself at the time.

His home country could indeed have put such help to good use, had it been available at the time. In the pre-war years, Belgian political diplomacy was faced with an impossible task. In 1839 a neutrality treaty had been enforced on the country because of its strategic position between France, Germany and England. In the early twentieth century, when the tensions between the great powers grew, this neutrality was changing into a death trap. On the one hand, isolation had to be prevented at all costs, since Belgium was dependent on Germany for its economy and on France for its culture – Sarton's father bought French wines from German merchants[12] – while England was the vital protector of its neutrality. On the other hand, exclusive alliances with any of the great powers were prohibited by the neutrality treaty.

After several decades of tension and a series of crises, keeping the balance became impossible when Germany confronted the Belgians with a request for free passage through their territory, in order that Germany might defend itself against the supposedly belligerent French. The request effectively cornered the Belgian government. Giving in would turn them into a German ally, with all due consequences. Not giving in would turn Belgian territory into a battlefield. A no-win situation, as the ensuing course of events has confirmed: in August 1914 Belgian neutrality was violated and the country was overrun by the German army.[13]

A year before the outbreak of the Great War, when all seemed not yet lost, Sarton attended the *Congrès mondial des Associations internationales* held in Ghent and Brussels in June 1913. In a paper entitled 'L'histoire de la science et l'organisation internationale' he argued that long-term peace could only be maintained if all nations strove for global unity.[14] Petty nationalism and

---

12 May Sarton, *I Knew a Phoenix* (n. 10), 12.

13 See, for instance, E.H. Kossmann, *The Low Countries: 1780–1940* (Oxford, 1988); Sophie de Schaepdrijver, *De Groote Oorlog: het Koninkrijk België tijdens de Eerste Wereldoorlog* (Amsterdam, 1997), esp. ch. 2.

14 The paper was published in the proceedings of the conference and in *La Vie Internationale*: G. Sarton, 'L'Histoire de la Science et l'Organisation Internationale', *La*

international rivalries based on culture or race were phenomena of the past. Those who still harboured such chauvinist and racist feelings were ignorant and narrow-minded, for they failed to read the signs of the time. Sarton had to acknowledge, however, that this message still had to be driven home to the leaders of the great powers and to the builders of cannons and army suppliers. Here, an enormous task lay ahead for international organisations and for the practitioners of a new profession, the *organisateurs internationaux*.

Sarton thus showed himself a supporter of the internationalist creed that was spreading rapidly in small European countries such as Belgium and the Netherlands in this period. The paper also makes clear that it was in this context that he found his destination in life. He saw himself as one of those international organisers who set themselves the task of facilitating the realisation of international unity, of helping to turn the world into a family of nations.[15]

The process of internationalisation was already well underway, Sarton believed. Now the main task was to stimulate it and to explain its background and operation. The main force behind it was science. Scientific research was the only truly progressive human activity. In its steady approach towards the truth, it could not fail, in the end, to unite all nations. Science knew no boundaries, no cultural differences, no racial prejudices. Its results were the product of the cooperative activity of humankind as a whole. What was true in science was true for all humans, and thus it could provide a solid foundation for international consensus in human affairs: '*science is the world's communal thought*'.[16] Moreover, in its alliance with technology, it would ensure prosperity and welfare for all.

This combination of Comtean positivism and Edwardian optimism was to remain Sarton's principal source of inspiration and the prime rationale for his life's work.[17] His outlook was challenged by two world wars that deeply affected him, and in later years he confessed to be 'not as cocksure as I used to be' about

---

Vie Internationale 4 (1913), 27–40. Sarton republished it in its original form in *Isis* 29 (1938), 311–25; references are to the latter version. This and all subsequent translations from the French are mine.

15 Ibid., 313. On Belgian internationalism of the period, see Daniel Laqua, Christophe Verbruggen and Gita Deneckere (eds), *Beyond Belgium: Encounters, Exchanges and Entanglements, 1900–1925*, special issue of *Revue Belge de Philologie et d'Histoire/Belgisch Tijdschrift voor Filologie en Geschiedenis* 90 (4) (2012). See also Pyenson and Verbruggen, 'Ego and the International' (n. 5), 62–3.

16 Ibid., 316, Sarton's italics.

17 For nineteenth-century positivism and ideas on progress in relation to the history of science, see Rachel Laudan, 'Histories of the Sciences and Their Uses: A Review to 1913', *History of Science* 31 (1993), 1–34. For the examples followed by Sarton (e.g. Paul Tannery), see also Lewis Pyenson, 'Prerogatives of Human Intellect: Historians of Science and the Promotion of Western Civilization', *History of Science* 31 (1993), 289–315.

Comte's brand of positivism.[18] But Sarton would cling to his conception of science as an international, rational, cumulative and progressive enterprise for the rest of his life.

In 1913, as indicated, he imagined his own contribution to bringing about international unity-via-science to consist in taking up the role of an 'international organiser'. In this capacity, his chosen field would be the history of science. Editing *Isis* was one way to fulfil his mission: its original subtitle was *Revue consacrée à l'histoire et à l'organisation de la science.* Another means to the end of promoting the cause of international unity was the writing of a synthetic history of science. It was useless, Sarton argued, to try to enforce international cooperation. The best results were to be expected from a convincing demonstration of the unifying and beneficial effects of science, that is, from showing the world how science had, from its beginning and throughout its history, been an activity to which people of different races, cultures and civilisations had contributed, an activity that knew no boundaries but was vitally dependent on international cooperation and that in its turn, by the insights it provided and the benefits derived from it, became a force of civilisation in its own right, until at last the whole world would, in peaceful cooperation, be engaged in its development.[19]

There was no German, English or French science – Sarton chose his examples carefully – and no individual had ever made a discovery entirely on his or her own. Not even nations could claim discoveries, for there were always traditions of research involved that transcended national and cultural boundaries. For the same reasons the unity of science, which according to Sarton followed necessarily from the unity of nature, had never been broken. Despite the increasing trend toward specialisation, the disciplines continued to be vitally dependent on each other. The history of science should therefore illustrate the progress of science as a unified whole and show how the growth of science had become more and more synonymous with the progress of civilisation and of humankind: '*thus the history of science is truly a history of the origin of culture and of human progress*'.[20] Showing the workings of science and its impact on world affairs, Sarton believed, provided the best means of convincing everyone of the desirability and necessity of worldwide unity and global solidarity. Or, as he would put it later in his *Introduction to the History of Science*: 'The unity of nature, the unity of knowledge, and the unity of mankind are but three aspects of a single reality'.[21]

---

18 Bern Dibner, 'Sarton Letters at the Burndy Library', *Isis* 75 (1984), 45–9: 46. How the German Violation of Belgian Neutrality in 1914 Affected Sarton Can Be Gleaned from G. Sarton, 'The Future of Belgium', *The Open Court* 29 (1915), 257–72.

19 Sarton, 'L'Histoire de la Science et l'Organisation Internationale' (n. 14), passim.

20 Ibid., 317, Sarton's italics.

21 Sarton, *Introduction* (n. 4), vol. 1, 29–32; vol. 3 pt. 1, 19–20: 19.

In his introduction to the first issue of *Isis* Sarton followed the same line of reasoning, stressing again that science was the only truly progressive and international activity that bound the peoples of all nations together: '[S]cience is the powerful peacemaker; it's the cement that unites the highest and most understanding minds of all nations, of all races, of all religions'.[22] The history of science had to catalogue and demonstrate the progress of science and thus the progress of humanity. In the scientistic vein that was not uncommon in this period, Sarton speculated that the study of the history of science might possibly even uncover the laws of scientific progress and the laws of human intelligence. *Isis* would finally, whenever necessary, denounce nationalistic prejudices and imperialist tendencies of any country or race.[23]

Isis also served the purposes of what Sarton came to call the New Humanism. Specialisation had driven apart the rapidly expanding sciences and the traditional, humanistic forms of learning, even to the point that the gulf between them threatened to obscure the unity of knowledge and thus became an impediment to human progress. Moreover, science could be used for 'sinister purposes', as in times of war, and this further alienated humanists from the scientific enterprise. There was, however, a dependable bridge between the sciences and the humanities: the history of science. By demonstrating the intrinsically human nature of scientific research, the history of science closed the gap between workers in seemingly disparate fields, thus satisfying what Sarton considered to be the main cultural need of his time:

> [S]cience must be tempered by humanity, and the best way of doing this is to explain its organic development, and to show all that was really great, beautiful and noble in [it].... That mixing of the historical with the scientific spirit, of life with knowledge, of beauty with truth, is the very essence of... the New Humanism.[24]

> The New Humanism is essentially the humanization of science or rather its re-integration with other elements of our culture.[25]

Both scientists and humanists were needed 'to go forward and prepare the dawn of a new age – the age of a New Humanism'.[26] It was also in the context of this neo-humanistic ideal that Sarton would go so far as to state, 'History itself is of no concern to us. The past does not interest us but for the future.... *History is but a method – not an aim!*'[27]

---

22 G. Sarton, 'L'Histoire de la Science', *Isis* 1 (1913), 3–46: 43–4.

23 Ibid., 37–8, 44.

24 G. Sarton, 'War and Civilization', *Isis* 2 (1919), 315–21: 319.

25 G. Sarton, 'Knowledge and charity', *Isis* 5 (1922), 5–19: 18.

26 G. Sarton, *The History of Science and the New Humanism* (New York, 1931), 72.

27 G. Sarton, *The Life of Science: Essays in the History of Civilization* (New York, 1948), 58, Sarton's italics.

Practical and economic reasons forced Sarton to leave Belgium in 1914. The German army commandeered parts of his house in Wondelgem, and the small government subsidy that helped him to publish *Isis* was discontinued after Belgium had capitulated. Via England, his wife's home country, where finding a suitable job proved to be impossible, Sarton sailed to the United States, attracted by its greater diversity of academic institutions. Here he was finally offered the position of research associate at the Carnegie Institution, while Harvard University furnished him with an office in the Widener Library, in exchange for a lecture series. Widener would remain Sarton's professional home base until his death in 1956. Fundamentally, his views on science and its history would never change and he would continue to defend them. In its most concise form, this defence was to take on the form of a theorem *cum* corollary:

> *Definition*. Science is systematized positive knowledge, or what has been taken as such at different ages and in different places.
>
> *Theorem*. The acquisition and systematization of positive knowledge are the only human activities which are truly cumulative and progressive.
>
> *Corollary*. The history of science is the only history which can illustrate the progress of mankind. In fact, progress has no definite and unquestionable meaning in other fields than the field of science.[28]

Sarton never lost his internationalist zeal either. For him, the history of science, the New Humanism and internationalism were inseparable in their political and moral implications. The New Humanism was to be 'the very armature of the international movement, – that slow but steady progress toward the political unification of the human family'.[29] Sarton made no bones about using *Isis* as a platform for expressing his political views. As I.B. Cohen put it, Sarton 'peppered *Isis* with personal editorials, many of which were like the newspaper columns of political pundits'.[30]

For Sarton, as for many intellectuals in the English-speaking world, science and democracy were closely linked, and the progress of science was therefore threatened by totalitarianism.[31] His abhorrence of the resurgence

---

28 G. Sarton, *The Study of the History of Science* (Cambridge, MA, 1936), 5.

29 G. Sarton, 'Preface to Volume IX', *Isis* 9 (1927), 226–33: 228–9.

30 I. Bernard Cohen, 'The *Isis* Crises and the Coming of Age of the History of Science Society', in Margaret W. Rossiter (ed.), *Catching up with the Vision: Essays on the Occasion of the 75th Anniversary of the Founding of the History of Science Society*. A supplement to *Isis* 90 (1999), 28–42: 29.

31 David Hollinger, 'The Defense of Democracy and Robert K. Merton's Formulation of the Scienctific Ethos', *Knowledge and Society: Studies in the Sociology of Culture Past and Present* 4 (1983), 1–15.

of political and scientific nationalism in the 1930s was the theme of several of his editorials. In one of these, for instance, Sarton defended democratic values by retelling a story from the Arabian Nights. The story, which no reader in the late 1930s could fail to read as an allusion to the situation in Germany, showed how people could be controlled and disciplined in every aspect of their lives by force-feeding them certain drugs. The result was loss of individual freedom, of conscience, of social life and of progress. This was, Sarton pointed out, what would inevitably happen even in highly civilised countries where the state was personified by its leader and where propaganda and coercion maintained the social order. Dictatorship might serve the ends of war, but it left the people poorer, weaker and dejected.[32] Without political freedom and freedom of thought, he wrote elsewhere, the progress of science became impossible.[33]

Although Sarton's radical convictions gave way to more moderate liberal values in his adult life, he still confessed to feel sympathy for the communist reaction against the increasing distance between the extremes of poverty and wealth in the capitalist West. Otherwise, he had little faith in the feasibility of the communist system, for in the end, it was the individual, not the community, that was at the core of social life: 'Restraint must come from within'.[34]

On the eve of the Second World War, in 1938, Sarton felt compelled to reprint in *Isis* his 1913 lecture on scientific internationalism and the unity of humankind discussed above.[35] Introducing the text, he stated that he would now, 25 years later, have worded certain passages differently, yet basically he still endorsed the same views, while his message had lost nothing of its relevance: only science could unite the peoples of all nations, and its history comprised the most noble aspects of the history of mankind.[36]

However, again like many American intellectuals, Sarton had exchanged his youthful pacifism for interventionism by this time: 'Happily the raid of Pearl

---

32 G. Sarton, 'Preface to Volume Twenty-Eight', *Isis* 28 (1938), 321–9.

33 G. Sarton, 'Third Preface to Volume XXXIV', *Isis* 34 (1943), 193–5. Joy Harvey has suggested that Sarton's left-wing views may have been the cause of a growing tension between him and Laurence J. Henderson, the other protagonist of the history of science at Harvard and a leading member of the right-wing Harvard Pareto circle: J. Harvey, 'History of Science, History and Science, and Natural Sciences: Undergraduate Teaching of the History of Science at Harvard, 1938–1970', in Rossiter, *Catching up with the Vision* (n. 30), 270–94, esp. 277–8.

34 G. Sarton, 'Preface to Volume XXVI. Confiteor', *Isis* 36 (1946), 89–91.

35 G. Sarton, 'Deuxième Préface au Volume Vingt Neuf. L'Histoire de la Science et l'Organisation Internationale', *Isis* 29 (1938), 311–25; see also note 14.

36 Sarton, 'Deuxième Préface au Volume Vingt Neuf', 312.

Harbor (7 Dec. 1941) awakened us . . . and the world was saved'.[37] During the war, Sarton published an undisguised outcry against Hitler in *Isis*, justifying his editorial by pointing out that there were misguided scientists amongst Hitler's followers, a shocking fact which affected German culture as a whole and for which historians of science would have to find an explanation.[38]

The war that made his home country suffer under German domination for the second time in the space of a few decades left Sarton dispirited, yet it did not change his deepest convictions. The war brought with it a 'fantastic abuse of science', he wrote in 1943, and this would inevitably cause 'a violent revulsion of feeling against it', yet this only made him plead with all the more urgency for his New Humanism and for the central role of the history of science in it.[39]

Unity and biohistory

It was the political uncertainty of the late 1930s that made Frans Verdoorn decide to leave his home country and move his publishing company to the United States. There was little sign of the romantic idealism that characterised the young Sarton in Frans Verdoorn. Born in 1906 as the son of a tobacco company director, scientific precociousness was his most conspicuous quality.[40] In his high school years he began to publish on the taxonomy of mosses and to correspond with the world's leading bryologists. He had reached the status of an international expert in the field before he finished his university education in biology in 1933. In the same year he founded the Chronica Botanica Company, which, beginning in 1935, published the *Chronica Botanica* series, an annual global census of plant science research.

As the initiative to start this series suggests, Verdoorn, like Sarton, was a scientific internationalist. He shared this conviction with many Dutch intellectuals of the period. Toward the end of the nineteenth century the Dutch, like the Belgians, had become well aware how precarious the situation of small countries, neutral or not, would be in case of a military conflict between the major European powers. In the second half of the nineteenth century the Dutch had invested much energy in strengthening their national culture and identity, for it was clear that the Netherlands could no longer defend its right

37 Cited in Van Oye, *George Sarton* (n. 10), 124. On Sarton's changed tone see Christoph Meinel, 'Sarton, Science, and the End of History', *Berichte zur Wissenschaftsgeschichte* 8 (1985), 173–9.

38 G. Sarton, 'Third Preface to Volume XXXV. Scripta Manent', *Isis* 35 (1944), 201–5.

39 Cited in James Herschberg, *James B. Conant: Harvard to Hiroshima and the Making of the Nuclear Age* (New York, 1993), 408–9.

40 See note 1 for biographical works on Verdoorn.

of an independent existence by military means.[41] Yet obviously it was doubtful if a strong national identity would suffice to prevent the country from being overrun by more powerful aggressors. Dutch intellectuals therefore stressed the need for active involvement in international cooperation and peacekeeping. Small countries, they felt, were particularly suited for this task, since they presented no military threat. The Dutch, given their international experience and polyglot capabilities, should take the lead in such initiatives and act as guides in international affairs.

An example of the fervour of the Dutch in this sphere is provided by their activity in international law, resulting in the building (with Carnegie funds) of the Peace Palace and the establishment of the permanent International Court of Justice in The Hague in the early twentieth century. (In a similar vein, Sarton suggested in 1915 that a Belgian city would be an ideal residence for the central offices of all international endeavours.[42]) Science provides an equally convincing example. Several Dutch scientists were to take a leading part in the promotion of international cooperation and the establishment of international scientific organisations. The role that H.A. Lorentz, the physicist, played in the League of Nations is the best known example. He took the lead in Dutch attempts to restore the relations between scientists from Germany and the Allied Powers in the interwar period. He was seconded in this by one of Frans Verdoorn's teachers in Utrecht, the botanist F.A.F.C. Went, one of the most active and driven internationalists among Dutch scientists.[43]

In the Netherlands, too, the history of science was put to the service of internationalism. In 1896, for instance, the medical history journal *Janus* was founded, which, according to its first editors, should contribute to the progress of medicine and to the promotion of welfare, peace and the concord of peoples. Like Sarton a few decades later, *Janus*'s editors did not propagandise the history of medicine and medical science as an end in itself. History was the handmaiden of science. They explained the two-faced Janus head emblem from which the journal derived its name as an illustration of their conviction that in order to ensure progress in the present and the future, one had to keep a constant eye on the past. There were, for instance, important methodological lessons to be learnt from how scientists had conducted their research in

---

41 For this 'cultural nationalism', see Jan Th.M. Bank, *Het Roemrijk Vaderland. Cultureel Nationalisme in Nederland in de Negentiende Eeuw* ('s-Gravenhage, 1990).

42 Sarton, 'Future of Belgium' (n. 18), 270.

43 For an overview focusing on the role of the sciences in this movement in the interwar period, see W. Otterspeer and J. Schuller tot Peursum-Meijer, *Wetenschap en Wereldvrede. De Koninklijke Akademie van Wetenschappen en het Herstel van de Internationale Wetenschap tijdens het Interbellum* (Amsterdam, 1997), ch. 1. For more on Dutch internationalist initiatives, see Chapter 5, this volume.

the past. To guarantee the steady growth of knowledge, every scientist should not only be aware of the discoveries made by his forerunners, but also of their errors. Science was a rational, progressive and cumulative activity, and its history constituted the foundation on which to build its future. This was why Janus looked forward and backward simultaneously.[44] Again, there is an obvious similarity to Sarton's conviction that the history of science was the history of 'man's quest for truth and gradual approach to it, of his struggle against error and unreason'.[45]

Went, the botanist, also propagated the idea that the history of science was important for the working scientist. In his teaching, he insisted that scientific research was not an impersonal, but an intrinsically human activity. Plants produced not a grain of botany, botany was the sum of plants and humans, he used to say. Therefore botanical knowledge could never be entirely separated from its makers, and thus the history of botany, which studied the development of human-plant relations over time, contributed to a better understanding of science. It was from Went that Frans Verdoorn inherited both his internationalism and his historical interest.[46] His work as a taxonomist, which naturally led him to the historical sources that formed the basis for nomenclature, further stimulated his predilection for the history of science.

Verdoorn's scientific internationalism was at the basis of the Chronica Botanica publishing company that he founded in 1933. His internationalist outlook is also evident from the papers he wrote on international scientific cooperation and from his role as secretary (from 1935 until 1953) of the International Union of Biological Sciences (IUBS).[47] Like Sarton, he saw the history of science as one of the means to further the cause of international scientific cooperation. A characteristic example of Verdoorn's internationally oriented historical activities is his *Index botanicorum* project, an ambitious venture with collaborators from all over the world that was intended to result

---

44 B. Theunissen, 'Journals of the History of Science in the Netherlands', in M. Beretta, C. Pogliani and P. Redondi (eds), *Journals and History of Science* (Florence, 1998), 197–210, esp. 202.

45 G. Sarton, 'An Institute for the History of Science', *Isis* 28 (1938), 7–17: 13.

46 F. Verdoorn, *Iter Biohistoricum. Een Excursie in de Grensgebieden tussen de Biologie en de Humaniora* ('s-Gravenhage, 1958), 7.

47 For instance, F. Verdoorn, 'International Cooperation among Herbaria and Taxonomists', *Chronica Botanica* 5 (1939), 140–42; idem, 'The Future of Biology in World Affairs', *Nature* 154 (1944), 595–9; idem, 'The International Plant Science Congresses. Their History and Aims', *Proceedings of the 7th International Botanical Congress, Stockholm* (Stockholm, 1950), 42–56. Verdoorn was involved in the establishment of several IUBS organisations, such as the Index Herbariorum, the International Association of Plant Taxonomists, and the International Biohistorical Commission.

in the publication of an international biographical encyclopaedia of botanists.[48] History for its own sake was not the main aim, as appears from Verdoorn's announcement of the project: 'I am preparing an "Index botanicorum", which will contain detailed biographies of all plant science workers, whose *curriculae vitae are of direct scientific interest*.'[49]

Realising that the prospects for his international company would be dim if Europe became a battlefield, Verdoorn emigrated to the United States in early 1940. Here, under the influence of Sarton, whom he befriended in the 1940s, Verdoorn began to set his historical ambitions higher and higher. In 1944 the shift in his interest from botany to history became manifest even in the *Chronica Botanica* series, which from then on had as its subtitle: *An International Collection of Studies in the Methodology and History of Biology and Agriculture*. In 1948, Verdoorn started the *Pallas* book series, a collection of reprints of classic scientific works. In 1950, at the Seventh International Botanical Congress in Stockholm, he successfully pleaded for the inclusion of a section for the history of botany in future congresses.[50] The concept of biohistory began to take shape in the late 1940s, and in 1951 Verdoorn was one of the founders of the Boston Biohistorical Club, of which Sarton was honorary chairman.[51] True to his entrepreneurial nature, Verdoorn also began to explore the possibilities of establishing a biohistorical institute at an American university, but he made little headway in this direction. The invitation from Utrecht, in 1953, therefore came as if on call.

Originally, the offer had been for a staff position at a new institute for the history of science of which the historian of science E.J. Dijksterhuis was to be the director and chair. It testifies again to Verdoorn's acute business sense that the ensuing negotiations with the university board resulted in Verdoorn being appointed as professor of biohistory in the Department of Biology, while Dijksterhuis ended up in a room in Verdoorn's new Biohistorical Institute.[52]

What was meant by this neologism – biohistory? For Verdoorn biohistory was not simply history of biology, but the exploration of the many borderlands between the life sciences and the humanities. Sarton's influence was at work

48 F. Verdoorn, 'The Index Botanicorum', *Chronica Botanica* 3 (1937), 335–6. The project was never finished. It probably collapsed under its own weight.

49 Ibid., 335, Verdoorn's italics.

50 Verdoorn, 'International Plant Science Congresses' (n. 47), 457.

51 F. Verdoorn, 'Medicine and Arts: Introduction to a Special Session on Medicine and the Humanities', *Proceedings of the XXIII Congress on the History of Medicine (London 2–9 September 1972)* pt. 2 (London, 1974), 865–9, esp. 869.

52 For details on the offer from Utrecht, see Klaas van Berkel, *Dijksterhuis* (Amsterdam, 1996), 440–43. The end result of the negotiations seems to have satisfied Dijksterhuis no less than it did Verdoorn.

Figure 9.2 Frans Verdoorn boating at Lake Sunapee, New Hampshire, summer 1944.

Source: Collection Institute for History and Philosophy of Science, Utrecht University. Reproduced with permission.

here, but Verdoorn also drew his inspiration from medical historians such as William Osler and Henri Sigerist, whose interests included art historical, literary and social aspects of the life sciences.[53] Although the idea of 'borderland research' was the starting point of Verdoorn's plans for a biohistorical institute, the subject matter of biohistory needed to be classified and fleshed out in much more detail, he felt. As he readily admitted, Verdoorn was prone to the 'Linnaean inclination' to classify, reclassify and reclassify again. After his return to Utrecht, he would spend the better part of his further career on this groundwork.

[53] Verdoorn, *Iter Biohistoricum* (n. 46), passim; Smit, 'Biografie tot Biohistorie' (n. 1), 11.

In 1964, after several revisions, he had expanded his classification of the biohistorical subject matter to the extent that its exposition filled 13 pages.[54] Shortly thereafter he completely revised the scheme again and redefined biohistory as 'the historical relationships of plants, animals and man in science, early medicine and culture'.[55] Its major categories included the 'formative period' (Palaeolithicum to medieval times), the history of the life sciences, the history of man's control over nature, and ethnobiological, literary, art historical and linguistic aspects. A few years later the category 'historical social biology' was added, modelled after Henri Sigerist's historical social medicine.[56] On this occasion, Verdoorn overhauled his system once again and decided to take what he called 'the primary human motivations or drives', such as thought, speech, visual expression, social relations etc., as the starting point for his system, designating the subdivisions as ever more narrowly defined 'colours of the biohistorical spectrum'.[57] The history of biology was merely a small line in this spectrum. According to Verdoorn, it did not need more attention: restricting oneself to the internal history of science would lead to 'a hardening of our brains and a clerk-like declension of our spirits', he once remarked.[58]

From this last quotation it is clear that Verdoorn had by then drifted far away from conventional conceptions of the history of science. Even so, there were still similarities with Sarton's approach. Whereas Sarton had stressed the unity of science, Verdoorn's efforts to refine his biohistorical classification were driven by a desire to illustrate the 'unity of biohistory'.[59] Another resemblance is their shared preoccupation with bibliography, indexing and encyclopaedic compilation. Like Sarton's, Verdoorn's historical projects were of a predominantly encyclopaedic nature. Finally, and most importantly, his internationalism remained one of the principal driving forces of Verdoorn's historical work. In his words: '[W]e cannot make the modern,

---

54 F. Verdoorn, 'Vademecum Biohistoricum', unpublished manuscript [Utrecht 1964], VP. For previous revisions see F. Verdoorn, 'Biohistory, its Aims and Scope', *Actes du VIIIème congrès internationale d'histoire des sciences (Florence 3–9 Septembre 1956)* pt. 2 (Florence, 1958), 762–9; idem, *Iter Biohistoricum* (n. 46), 10–13.

55 F. Verdoorn, 'From Botanical Biography towards Animal Iconology', *Acta Botanica Neerlandica* 15 (1969), 86–94: 89.

56 F. Verdoorn, 'De Plant in de Biohistorie', *Jaarboek der Koninklijke Nederlandse Botanische Vereniging over 1970* (1971), 29–84, esp. 71–4.

57 Ibid., 37–8.

58 Verdoorn, 'Medicine and Arts' (n. 51), 868.

59 Verdoorn, *Iter Biohistoricum* (n. 46), 10; idem, 'De Plant in de Biohistorie' (n. 56), 33, 83.

highly specialized biologist, with his limited outlook, more internationally minded . . . unless we stimulate research in and a better understanding of the untrodden borderlands between the humanities and the natural sciences'.[60]

Yet that same issue marks the point where Sarton and Verdoorn went their separate ways. The idealistic young Sarton had chosen the writing of a synthetic history of science as his mission in life. He considered it to be a humanitarian project: the history of science was a central element in his ideology of a new humanistic brotherhood of all nations. Sarton's optimism and ambitions were dampened by the experiences of a lifetime, but the core of his views on the role of science and its history remained intact. Verdoorn was of a younger generation. Although internationalism, as with many Dutch scientists born around 1900, had been second nature to him and had inspired his historical work, he did not have such high hopes for the history of science as an instrument for universal peace and understanding as Sarton had entertained. It was one thing to consider science to be an important factor in international relations and to derive humanistic lessons from its history, it was quite another to give the history of science central stage in human history or to see it as the harbinger of worldwide unity. Although he liked to quote from Sarton's 'humanistic' writings, Verdoorn never seriously proselytised for Sarton's New Humanism. In his Utrecht years he stated explicitly that it was an exaggeration to consider the history of science as the 'real' history of mankind.[61]

Verdoorn did not doubt that biologists and international scientific relations would profit from the results of his biohistorical programme in many ways, yet the programme itself, as an intellectual endeavour, also deserved to be studied for its own sake, as a part of cultural history. Biohistorical studies uncovered 'eternal values in our cultural history, values which – even though we often do not realise it – constitute an essential part of modern culture'. Therefore the question 'Cui bono?' with respect to such studies was out of place. Biohistory contributed to the understanding of human culture and needed no additional justification.[62] Put differently, while Sarton's interest was, ultimately, in science rather than in history, and in the future rather than the past, Verdoorn's was in history rather than in science, and in the past rather than the future.

---

60 Verdoorn, 'International Plant Science Congresses' (n. 47), 55.

61 F. Verdoorn, 'Problems of Botanical Historiography', *Archives Internationales d'Histoire des Sciences* 15 (1951), 448–57, esp. 451.

62 Verdoorn, 'Plant in de Biohistorie' (n. 56), 77.

By the same token, however, Verdoorn's work is the more vulnerable to the criticism levelled at Sarton's writings by Arnold Thackray and Robert Merton.[63]

Ends and means

Some comments on this criticism are in order here. Important though Sarton was for the professionalization of the history of science, Thackray and Merton have argued that he nevertheless failed to provide the discipline with a cognitive identity, a clearly defined problematic that could have guided research in the field. Sarton was no founder of schools. Towards the end of his life his historical approach was outdated, and his New Humanism met with little response. With respect to its research problems and methodological approaches, the history of science was shaped by others, such as Alexandre Koyré in the United States and Herbert Butterfield in England. Sarton's contribution consisted in providing research aids: surveys, guides, bibliographies, encyclopaedic introductions and of course a journal. The paradox of George Sarton, Thackray and Merton concluded, is that he contributed to laying the foundations for the history of science as a professional discipline without defining a specific conceptual and analytical approach that was indispensable for it to become fully articulated and coherent.

Obviously, this criticism is to the point in retrospect: Sarton's approach to the history of science has not survived him.[64] Yet seen from Sarton's own perspective on what the history of science should be and should do, I think it misses the point of his mission. For Sarton, means and ends were perfectly attuned to each other. In view of his New Humanism, the ends were to document the gradual growth of positive knowledge, to take stock of what scientists had accomplished from ancient times to the present, to demonstrate that science was at the core of human progress and civilisation and, finally, to prove the transnational, transracial and transcultural nature of the scientific enterprise. To achieve these goals, Sarton decided to work in a strictly chronological order, to document carefully 'who did what where and when (first)' and to show how the torch of progress was constantly passed on from one culture or nation to another. The compilation of bibliographies was indispensable for this project, and Sarton's use of numerical analyses, tables and graphs – for instance to compare the contributions of different countries or civilisations – must also be seen in this light.[65]

---

63 Thackray and Merton, 'On Discipline Building' (n. 5).

64 Yet for a more positive evaluation of his encyclopaedic work, see Pyenson, 'Inventory as a Route to Understanding' (n. 5).

65 Compare Thackray and Merton, 'On Discipline Building' (n. 5), 491–3.

The result, to be sure, was an inventory of knowledge production, as the five-volume *Introduction to the History of Science* amply illustrates. Yet the compilation work was not intended merely to lay the groundwork for something different, for a programme of research that for some reason or other was never formulated. Sarton's bibliographic and encyclopaedic work formed the core of the positivistic project that he had embarked upon. The synthetic works that he planned were not to be essentially different in approach from the volumes that he did publish. The synthetic writings had to wait, however, until all the relevant data on the development of science in a certain period had been brought together in manageable inventories. The next step would then be scrupulously to compare, weigh and evaluate the data with respect to their truth or falsity and importance. Thus scrutinised and selected, the data could then be used to write an exhaustive – synthetic – work on the period that showed how positive knowledge had grown, how humanity had gradually progressed toward the truth.[66]

It is clear that Sarton had hopelessly underestimated the daunting task that he had set himself in his youthful enthusiasm. Even the *Introduction to the History of Science*, which he had initially hoped to complete in less than a decade, proved to be too big an undertaking for one man to finish in a lifetime. Consequently, the projected synthetic works remained unwritten. Yet this does not alter the fact that the books that he did finish were no mere bibliographic and encyclopaedic tools for a programme of research yet to be undertaken. They were part and parcel of a single project. So there is no paradox here.

Moreover, the task of Sartonian historians of science comprised even more than writing synthetic works. Apart from analysts and historians they should, according to Sarton, also be humanists who helped to humanise science and bring about the New Humanism.[67] Historians of science were ideally suited for this work, since their daily work as annalists 'prove[d] inductively the unity of knowledge and the unity of mankind'.[68] Thus Sarton's own 'humanistic' writings were again in perfect harmony with the goals he had formulated.

Frans Verdoorn's case, on the other hand, was different. Once he had begun his biohistorical project, he was no longer principally engaged in a humanitarian project but tried to establish a new field of research that he claimed to be of

---

66 In an essay on the history of twentieth-century science, Sarton indicated in detail how historians might proceed to get from inventory to synthesis: G. Sarton, 'Remarks Concerning the History of Twentieth-Century science', in Dorothy Stimson, *Sarton on the History of Science: Essays by George Sarton* (Cambridge, MA, 1962), 355–63.

67 G. Sarton, 'An Institute for the History of Science', in Sarton, *The Life of Science* (n. 27), 169–74, esp. 172.

68 Cited in Thackray and Merton, 'Sarton' (n. 10), 109.

intrinsic cultural interest and for which he needed a costly institute with an interdisciplinary staff and a well-stocked library. In his case the question of the cognitive identity of the new field might rightfully have been asked from the beginning. It was not asked at first, but when it finally was, at Verdoorn's retirement, there was no answer to be found in his writings.

What Verdoorn envisaged was an institute for the history of the relationships between plants, animals and humans in science, applied science, social relations of science, art, literature, linguistics and music. Naturally there are endless possibilities for research here, and studies that fit into Verdoorn's scheme are undertaken all the time. Verdoorn would have revelled, for instance, in the volume edited by Jardine, Secord and Spary, *The Cultures of Natural History*.[69] Because of the broadly cultural perspective taken by the authors and the interdisciplinary range of topics they cover, it can be taken as an apt illustration of the kind of studies that Verdoorn had in mind. Yet the crucial question is whether there are compelling reasons for merging all the 'biohistorical' fields of investigation into a new and independent historical research programme requiring its own institute and staff. Do the relations between plants, animals and humans represent a field that is so unique, do they ask for a methodology that is so specific, that they cannot be properly studied as part of research programmes in cultural history, history of science, art history, history of music, etc.? Verdoorn had, in a Sartonian vein, argued that the borderlands between biology and the humanities should be seen as a unity, yet the onus was on him to explicate the nature of this unity and to demonstrate what it was that bound the borderlands together. For if one does not want every 'borderland' between two fields to lay claim to discipline status, there should be something more than its subject matter to justify its independent status.

Verdoorn's numerous expositions of biohistory provide no justification of this kind. He failed to explicate the cognitive profile of his field, and in his case, this implied that his design for a new discipline remained an empty shell. Verdoorn's biohistorical system may have formed a unity from a taxonomic point of view, yet it was not a natural system in the Linnaean sense, but an artificial contrivance composed of subjects that did not form a generic unity.[70]

Still, without Verdoorn's trailblazing activities there would, in all probability, never have been a programme in the history of biology in Utrecht – the only line in the biohistorical spectrum that survived his retirement. One cannot help

69 N. Jardine, J.A. Secord and E.C. Spary (eds), *Cultures of Natural History* (Cambridge, 1996).

70 For more details on Verdoorn's classification of the biohistorical subject matter, see the works in note 8.

wondering, however, what it was that earned him his professorship at a biology department in the first place. And what role did 'Sartonian' ideas play in this respect? Had such ideas not been largely discredited by the early 1950s? For this we have to return to Sarton once more.

Humanising science

On the basis of a detailed, though rather unsympathetic account of Sarton's views, Tore Frängsmyr has questioned Thackray and Merton's explanation of why Sarton founded no school.[71] Frängsmyr suggested that what must have put historians off from the start was Sarton's New Humanism. In Frängsmyr's view, this ideology remained a castle in the air and had overtones that more sober minds must have found difficult to swallow. Likewise, Sarton's ideas on unity and synthesis struck Frängsmyr as hollow and essentially vague.[72] This interpretation, however, does not take the universalistic and internationalistic ideology into account that lay at the basis of Sarton's New Humanism and that he was far from alone in upholding.

As several contributions to the present volume amply show, the ideology of an international brotherhood of scientists and of the unity of science was highly topical in intellectual circles in the interwar years. True, Sarton's American patrons never considered his far-reaching claims for the central role of the history of science in furthering this ideology of sufficient urgency to sponsor his disciplinary ambitions.[73] However, as Thackray has shown, many of the early supporters of *Isis* and of the History of Science Society had European intellectual roots and did share Sarton's progressivism and internationalism. The history of science, for them, was a record of the successes of science, and thus functioned as an argument on its behalf. Likewise, they agreed with Sarton that the history of science could be put to the service of bridging the widening gap between the sciences and the humanities – a notion that had already gained currency in the context of the late nineteenth-century debate on overspecialisation in science.[74]

After the Second World War, the idea that the history of science, as a humanising force, helped to realise the goals of scientific internationalism lingered on until well into the 1950s. UNESCO, the United Nations' cultural and scientific branch, officially endorsed the notion of the supporting role of

---

71 Frängsmyr, 'Science or History' (n. 5).

72 Ibid., 111 and 128.

73 Sarton had made frequent pleas for the foundation of an institute devoted to the history of science. For example, Sarton, 'An Institute for the History of Science' (n. 45).

74 Thackray, 'Pre-History of an Academic Discipline' (n. 9).

the history of science in furthering UNESCO's creed of scientific humanism – a term coined by Julian Huxley.[75] At the instigation of Joseph Needham, director of UNESCO's natural sciences division, all international scientific congresses were requested to establish sections for the history of the science with which the congress was concerned.[76] It was also in this context that Frans Verdoorn successfully pleaded for the inclusion of a historical section in all future international botanical conferences.[77]

Similarly, the notion of the history of science as an instrument for bridging the gap between the sciences and the humanities had lost nothing of its appeal in the mid-twentieth century. '[It] is more than ever essential – if humanistic values are to be retained in general education – that the sciences be approached historically', wrote History of Science Society president John F. Fulton in 1947. In his view, bringing the sciences and the humanities together was 'one of the first problems of a modern university'.[78] His British opposite number, Charles Singer, stated in his 1948 presidential address to the British Society for the History of Science that a 'new humanism' was needed to heal disruptive intellectual discontinuities.[79] Similar expressions can be found in the works of Henry Guerlac, Herbert Butterfield and A. Rupert Hall.[80] In the Netherlands, the view that the history of science could bridge the gap between science and the humanities found another indomitable supporter in E.J. Dijksterhuis. The water that separated them was deep, Dijksterhuis told his students on the occasion of his inauguration in 1953, yet there was a ferry between the sciences and the humanities, the history of science, and he would be their ferryman.[81]

---

75 Julian Huxley, *What Dare I Think: The Challenge of Modern Science to Human Action and Belief* (London, 1931), 149–77. In its first declaration on race (1950), UNESCO also proclaimed the 'unity of man'; see Donna Haraway, *Primate Visions: Gender, Race and Nature in the World of Modern Science* (New York, 1989), 197–203. See also Chapter 3, this volume.

76 J. Needham, 'UNESCO and the History of Science', *Archives Internationales d'Histoire des Sciences* 1 (1947), 3–4.

77 Verdoorn, 'Problems of Botanical Historiography' (n. 62), 456–7. On Needham and the history of science in England, see Anna-K Mayer, 'Setting up a Discipline: Conflicting Agendas of the Cambridge History of Science Committee, 1936–1950', *Studies in the History and Philosophy of Science* 31 (2000), 665–89.

78 Quoted in Thomas P. Gariepy, 'John Farquhar Fulton and the History of Science Society', in Rossiter, *Catching Up with the Vision* (n. 30), 7–27: 13–14.

79 Charles Singer, 'The Role of the History of Science', reprinted in *British Journal of the History of Science* 30 (1997), 71–3, esp. 73.

80 Frängsmyr, 'Science or History' (n. 5), 129–30.

81 Van Berkel, *Dijksterhuis* (n. 52), 426–30 and 440–43.

Finally, Sarton's view of science as a progressive, rational and essentially autonomous enterprise cannot be held responsible for the lack of *Nachwuchs* for his historical approach, for despite the challenge to the notion of autonomy by the Bernalists in the 1930s, it remained the dominant view until the 1960s. It was also endorsed by the internalist and intellectualist approach to the history of science, personified by Koyré, that gained prominence in the 1940s.[82]

Sarton's idea of the crucial importance of the history of science for science, society and world peace was undoubtedly seen by many as too grandiose to be realistic; even the support of his patrons was moral rather than material, as illustrated by James B. Conant's granting him a programme at Harvard yet denying him the funds to build it up.[83] Yet it was not the progressivism, internationalism and humanism at the core of his convictions that made him an eccentric or his work an oddity. It was mainly, I would suggest, Sarton's encyclopaedic approach, his conception of the historian of science as an analyst and of the history of science as 'at once history and science, but . . . primarily science', that came to be seen as outdated and unattractive among historians of science.[84] As Anna-K. Mayer has shown, already for Needham and his 1930s Cambridge circle, history came first, not science.[85] And Frans Verdoorn may well have had Sarton's research method and conception of the field in mind when he quipped that narrowly restricting one's attention to the history of science was for clerks, not for biohistorians.[86] For Verdoorn, biohistory was not primarily, not even partly science, but cultural history.

It was probably not this insight, however, that gained Verdoorn the invitation for a position at the Utrecht biology department. It must rather have been the – in the early 1950s – perfectly timed way in which he presented his field: as an instrument of scientific internationalism on the one hand and

---

82 David Hollinger, 'Free Enterprise and Free Inquiry: The Emergence of Laissez-Faire Commutarianism in the Ideology of Science in the United States', *New Literary History* 21/4 (1990), 897–919. Hollinger argues that Thomas Kuhn's *Structure of Scientific Revolutions* only reaffirmed the faith in the autonomy of knowledge (see p. 913), while Steve Fuller even contends that 'Kuhn's book constituted . . . less a revolt against positivism than a continuation of positivism by other means'. See S. Fuller, *Thomas Kuhn: A Philosophical History for our Times* (Chicago, 2000), 287. Mayer has argued that the historians who controlled the Cambridge history of science scene after 1945 had an even more positivistic outlook on science than the scientists who preceded them (Joseph Needham among them). See Mayer, 'Setting up a Discipline' (n. 77). For a similar argument with regard to Koyré, see Frängsmyr, 'Science or History' (n. 5), 138–41.

83 Herschberg, *James B. Conant* (n. 39), 408.

84 G. Sarton, 'The Teaching of the History of Science', *Isis* 13 (1929–1930), 272–97: 273.

85 Mayer, 'Setting up a Discipline' (n. 77).

86 Verdoorn, 'Medicine and Arts' (n. 51).

as a bridge between the sciences and the humanities on the other. Thus, it was the 'Sartonian' part of Verdoorn's agenda for the history of science that played a major role in the institutionalisation of the history of science in the Netherlands.

Now, some 35 years after I was first attracted by the humanistic atmosphere in Verdoorn's institute, little is left of his legacy. There is one aspect however – the history of science as a humanising force – that continues to be of vital importance for the professional survival of historians of science in the Netherlands. As in Dijksterhuis's and Verdoorn's times, most of them are employed by science departments, and Sartonian overtones can still be heard in their rhetoric to defend their discipline.

Chapter 10

The unity of knowledge and the diversity of knowers

Science as an agent of cultural integration in the United States between the two world wars

David A. Hollinger

Seated across from a sceptical John Dewey, exasperated over his own apparent inability to explain logical empiricism and to persuade Dewey of the doctrinal ecumenism of the *International Encyclopedia of Unified Science*, Otto Neurath stood up abruptly, raised his right hand as if taking an oath and declared vehemently, 'I swear we do not believe in atomic propositions.' Whether Neurath's left hand grasped a copy of the Bible or of Isaac Newton's *Principia*, or only a wisp of startled air, we do know that Dewey was charmed by the gesture. According to Ernest Nagel, who, along with Sidney Hook, witnessed this conversation, Dewey immediately withdrew his resistance to Neurath's invitation that he write an article for the *Encyclopedia* and forthwith presented the agitated Neurath with a drink.[1]

The *International Encyclopedia of Unified Science* was the great dream of the scientific intellectuals who had emigrated from Hitler's Europe and of those Americans who welcomed Neurath and his fellow émigrés as instruments in the creation in the United States of a more secular, science-based culture, society and politics. Although the professional literature on twentieth-century American history has long since attended with great care and detail to the influence of Thomas Mann and Hannah Arendt among literary intellectuals and to the role of the émigré physicists Leo Szilard and others in the saga of the atomic bomb, historians have been less attentive to the philosophical dimensions of the transformation of American intellectual life by Adolf Hitler. To be sure, we have studies of the logical positivist movement and its impact on the long-term development of the discipline of philosophy. But the grandeur of the 'Unity

---

1 'Contribution of Ernest Nagel', in Corliss Lamont (ed.), *Dialogue on John Dewey* (New York, 1959), 10–13. The most informative and cogent account of the origins of the *Encyclopedia* and of the thinking of its founders is an essay published by Morris himself late in his career: Charles Morris, 'On the History of the *International Encyclopedia of Unified Science*', *Synthese* 12 (1961), 517–21. On the fate of the Unity of Science Movement in the US in the next two decades, see Chapter 11, this volume.

of Science Movement' – according to which the philosophical implications of all of modern knowledge worthy of the name would be incorporated and made available – has been largely lost from view, in part because this movement proved to be so spectacularly a dead end. But the path to this dead end tells us much about the intellectual life of the most science-inspired intellectuals in the United States during the middle decades of the twentieth century. The intellectuals who trod that path enacted the positive role that universalist ideas of knowledge could play in the diminution of ethno-racial and religious bias in American life. But the story of these intellectuals also reveals a parochial refusal on the part of the émigrés and even many of their American admirers to appreciate the range and depth of the scientific universalism that had already grown from indigenous American roots.

Hence the mid-1930s episode in Dewey's living room on Morningside Heights can introduce us to an important and engaging story. The conversation between Neurath and Dewey reveals a number of key elements of the history of unified science in the United States during the interwar years.

All four of the people in the room were identified politically with the socialist left, broadly considered. Ethno-religiously, Neurath, Nagel and Hook were freethinking Jews. Dewey was a lapsed Congregationalist well known for his lack of anti-Semitic bias at a time when discrimination against Jews was widespread among academics from Dewey's American-born, Anglo-Protestant background. Neurath was an émigré from Hitler's Europe then living in the Netherlands, Nagel had immigrated to the United States as a child and Hook was an American-born child of East European immigrants. This foursome thus displayed in microcosm the social character of the larger collection of people who took the most initiative in the United States to advance unified science in the 1930s. More Jewish, more leftist, more immigrant and more secular than the bulk of scientists and philosophers, the intellectuals who invested the most spiritual capital in what was called 'the scientific attitude' were strikingly representative of interwar American academia, including its scientific establishment. This 'attitude' had the greatest appeal to intellectuals who, whatever their own backgrounds, favoured an American future that would be more Jewish, more leftist, more immigrant and more secular than the American past.[2]

But the little incident in Dewey's living room is an emblem for more than the social, political and ethnoreligious demographics of our larger story. The incident also puts before us the intellectual constituencies and orientations that define this story. Neurath spoke for the aggressive and iconoclastic Vienna-centred movement of philosophers who embraced the concept of unified science with such ferocity and persistence that even in our twenty-first century the concept remains primarily associated with them. They were the new kids on the philosophical

---

[2] For the ethnoreligious demography of American intellectual life and its consequences, see David A. Hollinger, *Science, Jews, and Secular Culture: Studies in Mid-Twentieth-Century American Intellectual History* (Princeton, 1996).

block and made quite a production of announcing their arrival and willingness to fight. Dewey spoke for a well-established tradition of science-centred, publicly engaged philosophy that took the unity of science for granted. Dewey, who had spent more than 30 years writing on behalf of a cultural programme organised around scientific method, did not need Neurath to tell him that science was of a piece and that it was vital to make the world more scientific. In a reversal of the roles long standard on the trans-Atlantic intellectual stage, where excited young American thinkers displayed their naïve enthusiasms before a gallery of experienced and world-weary European savants, here was the seventy-six-year-old American Dewey playing the part of the somewhat jaded and sceptical senior host, concerned to put at ease the newest, most breathless member of the great seminar of the West.[3]

Dewey's scepticism reflected an awareness that what Neurath was selling as a universalist was patently particular. The 'we' on behalf of whom Neurath took his mock oath consisted, of course, of the philosophers of the Vienna Circle, who had allowed themselves to be popularly associated with the view that most of what American philosophers wrote was not only mistaken but, technically speaking, downright meaningless. The great Dewey himself was suspect. Pleading for Dewey within the private councils of the logical empiricists became a standard routine of Charles W. Morris, the young American disciple of George Herbert Mead who joined Neurath and Rudolf Carnap as co-editors of the *Encyclopedia*. Morris also spoke out in public. In a plaintive footnote in a 1935 article in the new journal *Philosophy of Science*, Morris objected to what he described as the all too common practice among 'scientific philosophers' of 'looking down on Dewey'. Morris called upon his collaborators to recognise that Dewey had 'accepted science to the hilt', understood science's 'cultural implications' and was massively engaged with the society in which science was obliged to operate.[4]

Dewey understood an important truth about Neurath's Unity of Science Movement that we today are inclined to forget as we focus on Neurath's flashes of philosophical wisdom, as well as on the doctrinal differences with Carnap that make Neurath a more distinctive thinker than he has sometimes been credited with being. The movement for unity, as led by Neurath in partnership with Carnap, was considerably more sectarian than its public language allowed. Sectarianism sometimes comes dressed as inclusiveness. The project for unified science, as presented to Dewey by Neurath, entailed a narrowing, rather than

---

3 This perspective on Otto Neurath and his place in the history of ideas differs considerably from that displayed in Nancy Cartwright, Jordi Cat, Lola Fleck and Thomas Uebel, *Otto Neurath: Philosophy Between Science and Politics* (Cambridge, 1996). For my own misgivings about this book, see my review of it in *American Historical Review* 105 (2000), 637–8.

4 Charles W. Morris, 'Philosophy of Science and the Science of Philosophy', *Philosophy of Science* 2 (1935), 280–81.

a broadening, of the movement for a more scientific civilisation that Dewey had led in the United States since before the First World War. At the end of that war Dewey experienced a triumphal lecture tour in China, coinciding with the May 4 Movement there, and presented in Japan the lectures published as *Reconstruction in Philosophy*, a book emblematic of the truly global scope of Dewey's programme and of his manifest ability to inspire audiences well beyond Neurath's reach.[5]

On just what terms was this narrowing enacted by the somewhat parochially European Neurath and his group? What potential forces for unified science did these terms leave out? What features of the larger movement for a science-centred culture in the United States enable us to sharpen our perspective on Neurath and his project? What did Dewey actually say when he did write for the first instalment of Neurath's *Encyclopedia* as published in 1938? What happened after 1938 to the relationship, such as it was, between Dewey's tradition of scientific thought and that of the encyclopaedists? These questions frame this article.

In order to cast into bold relief the rather selective approach that Neurath, even at his most inclusive, took to the *Encyclopedia*, we can look at the more capacious inventory of possibilities displayed at the same historical moment in the pages of *Philosophy of Science*. This journal, founded in 1934, was a major forum for the articulation of logical empiricism and was an enterprise promoted by many of the same people active in the *Encyclopedia*. Carnap and Herbert Feigl were both on the editorial board of the journal and contributed the first two articles in its opening issue. Yet the journal was much more inclusive than the *Encyclopedia*, especially in its greater receptivity to social science, other than psychology of the most physiological variety, and in its casual acceptance of Marxism as a science like any other.

Between 1934 and 1938 the journal published a number of contributions by social scientists, including Robert K. Merton's later-to-be-famous essay, 'Science and the social order'.[6] Most of the social scientific contributions were aggressively engaged in the endeavour to make the whole society more scientific. The economist and popular science writer Joseph Mayer contributed several articles along this line, as did the sociologist George A. Lundberg, later well known for his book *Can Science Save Us?* (1947). These writers, like countless social scientists and others in the United States in the 1930s, treated social

---

[5] John Dewey, *Reconstruction in Philosophy* (New York, 1919). For John Dewey's appeal in Asia, see Robert Westbrook, *John Dewey and American Democracy* (Ithaca, 1991), 240–52, where the limits of Dewey's appeal in the specific context of Chinese revolutionary politics are also explained.

[6] Robert K. Merton, 'Science and the Social Order', *Philosophy of Science* 5 (1938), 321–37.

science as a vital component of the effort to make society more scientific.[7] Nowhere, many of them would have said, is the unity of science more evident than in the use of the scientific method in the social, political and economic realms apparent in American departments of social science and in such social science journals as *American Journal of Sociology*, *Political Science Quarterly*, *American Political Science Review*, *American Economic Review* and *American Sociological Review*. Neurath himself occasionally spoke in this idiom and even within the pages of *Philosophy of Science*. In an April 1937 article reporting on plans for the *Encyclopedia*, Neurath himself allowed that the study of human beings and the study of the stars should be approached with the 'same logical techniques and with the same dispassionateness'.[8] Now, this is exactly what the American social scientists of the era thought they were already doing and they proclaimed it *ad nauseam*. To prove the seriousness of their endeavour, the American social scientists and their allies among philosophers had their own international encyclopaedia, *The Encyclopedia of the Social Sciences*, just then being published to great acclaim.[9]

But the vast social scientific establishment of American academia did not get into Neurath's *Encyclopedia*, nor even onto its advisory committee, despite Morris's constant agitating. Neurath's understanding of what American social scientists were doing appears to have been limited and Carnap too had little use for them. George A. Reisch, the closest student of the relationships between Neurath, Carnap and the other encyclopaedists, reported that it was Neurath, however, who was making most of the decisions and that Neurath was quite insistent upon having his own way.[10] To be sure, Chicago sociologist Louis Wirth was commissioned to do an article on 'The Sociology of Science', which

---

[7] This preoccupation and the internal divisions of this social scientific enthusiasm have been the subjects of extensive scholarly attention in recent years. Representative works include John Jordan, *Machine-Age Ideology: Social Engineering & American Liberalism, 1911–1919* (Chapel Hill, 1994); and Mark C. Smith, *Social Science in the Crucible: The American Debate Over Objectivity and Purpose, 1918–1941* (Durham, 1994). The most comprehensive study to date is Andrew Jewett, *Science, Democracy, and the American University: From the Civil War to the Cold War* (Cambridge, 2012).

[8] Otto Neurath, 'Unified Science and Its Encyclopedia', *Philosophy of Science* 5 (1937), 265–77, esp. 273–74.

[9] *The Encyclopedia of the Social Sciences* (12 vols, New York, 1933–1937) was a prodigious endeavour brought to successful completion by Alvin Johnson. This encyclopaedia is a much more important episode in the history of thought than *The International Encyclopedia of Unified Science* yet has attracted much less attention from historians than the abortive enterprise led by Neurath.

[10] See George A. Reisch, 'Planning Science: Otto Neurath and the *International Encyclopedia of Unified Science*', *British Journal for History of Science* 27 (1994), 153–75.

never appeared for reasons not known to me. But neither Wirth nor any of the Chicago sociologists were ever listed as members of the advisory committees to the unified science endeavour. So far as I have been able to determine, the advisory committees never included any American political scientists (although Charles Merriam, T.V. Smith and Harold Lasswell would have been obvious possibilities), never included any American economists (although Wesley Mitchell and George Soule would have been available) and never included any American sociologists of any persuasion. As for anthropologists, only the name of Edwin Sapir surfaces and then for only a short time in 1939. Behavioural psychologists Clark Hull, Karl Lashley and Edward Tolman, as well the Vienna émigré Egon Brunswik, did become part of the enterprise, along with the physicist Percy Bridgman and the human biologist H.S. Jennings.

Astonishingly, Neurath announced in 1937 that he himself would take on the task of writing the article on 'The Foundations of Social Science'.[11] When that curiously eclectic article, now remembered chiefly for its eloquent conclusion invoking Neurath's famous boat at sea, eventually appeared in 1944, Carnap objected that he had not been consulted according to normal editorial processes, and Nagel declared the piece to be a 'black eye' on the reputation of the *Encyclopedia*.[12] When Neurath died in 1945, even his idiosyncratic efforts to bring non-physiological social science into the unified science project were lost. What did survive was physiological psychology; the names of Brunswik, Hull and Tolman were always there, and they continued to be listed as deceased advisers in printings of the *Encyclopedia* long after they, themselves, were dead, generating a steady stream of knowing comments from sceptics to the effect that, beneath the false slogans of 'logic without metaphysics', there lurked a decidedly materialistic metaphysics.[13]

If the unwillingness of Neurath and Carnap to embrace mainstream American social science is the most striking example of the particularity of their project, their refusal to count Marxism as a science like any other is less surprising. It is

---

[11] [Otto Neurath, unsigned], 'Symposium of Unified Science', *Philosophy of Science* 4 (1937), 496–8.

[12] Otto Neurath, *Foundations of the Social Sciences* (Chicago, 1944), published as volume 2, no. 1, of the *International Encyclopedia of Unified Science*. The flyleaf of this pamphlet contains, after the standard list of members of the 'Advisory Committee' and 'Committee of Organization', this notation, in italics: '*Owing to special circumstances, Rudolf Carnap did not share the editorial responsibility for this monograph*.' Ernest Nagel characterised Neurath's contribution as a 'black eye' in a private letter to Charles Morris in November 1944, quoted in George A. Reisch, 'Disunity in the International Encyclopedia of Unified Science' (paper presented to the Conference on Logical Empiricism in North America, 2000).

[13] Ernest Nagel titled an important volume of essays *Logic without Metaphysics, and Other Essays in the Philosophy of Science* (Glencoe, 1956).

nonetheless highly pertinent to any effort to locate the distinctive path taken by the Unity of Science Movement through the spacious American discursive domain. It was difficult in the 1930s to extract Marxist ideas from the specifically Bolshevik project as institutionalised in the Comintern, about which many proponents of 'the scientific attitude' were increasingly dubious. Yet the pages of *Philosophy of Science* show the credibility of Marxist science in circles very close to Neurath and Carnap and during both the Third Period and the Popular Front. The first book review published in this journal's inaugural issue was a laudatory review of a new Soviet edition of Friedrich Engels's *Dialektik der Natur*. This review, published in the Third Period, casually mentioned the 'discovery' of dialectical materialism as an event in the history of scientific knowledge. It was written by Dutch émigré and journal editorial board member Dirk Struik, an MIT mathematician who had been elected to the American Academy of Arts and Sciences in 1930.[14] The second issue of the journal ran an article from the great geneticist H.J. Muller, also a member of the journal's editorial board. This piece was an address to the Soviet Academy, opening with a quotation from V.I. Lenin, delivered shortly before Muller's conflicts with Trofim Denisovich Lysenko led to his withdrawal from Moscow-led political action into a more democratic style of socialism.[15]

Philosophy of Science also included a number of articles by William Malisoff, its pluralistic editor, who said in his opening editorial that he welcomed all shades of philosophical opinion except the reactionary.[16] Malisoff ardently supported Neurath's *Encyclopedia*, joining the advisory committee sometime after 1938, along with the Berkeley physicist Victor Lenzen and the *New York Times*' scientific correspondent Waldemar Kaempffert, who was a relative of Neurath's. I suspect Malisoff would not have been added had Neurath and Carnap known a fact about him that was revealed more than a half-century later, when the US government released the transcripts of Soviet intelligence cables from the 1940s sent in a code that was cracked by American counter-intelligence agents. Malisoff, a native of Russia who had emigrated as a child, was a KGB agent for a number of years; in 1943 he even tried to get the KGB to contribute capital to his manufacturing business. As late as 1945 Malisoff offered to find out for the Soviets whatever they wanted to know about what was going on at the University of Chicago and offered to help influence a faculty appointment

14 Dirk Jan Struik, review of F. Engels, *Dialektik der Natur*, ed. D. Raisanov, *Philosophy of Science* 1 (1934), 122–3.

15 H.J. Muller, 'The Views of Haeckel in the Light of Genetics', *Philosophy of Science* 1 (1934), 313–22.

16 William Malisoff, 'What is Philosophy of Science?', *Philosophy of Science* 1 (1934), 123.

relevant to the metallurgical laboratory.[17] Malisoff died in 1947, still resolute, so far as we know, in his commitment to Stalin.

I have been using the contents of Malisoff's journal to indicate some of the possibilities for unified science that enable us to see the specificity of the version espoused by Neurath and Carnap. I want now to turn to a series of earlier efforts in the United States to advance the cause of a science-centred way of life. I have invoked Dewey as a leader of these efforts, and I will be mentioning him again. But the main point I want to develop about American efforts to promote 'the scientific attitude' between 1919 and the mid-1930s has to do with the relationship between the ethnic diversification of the community of knowers and the ideal of the epistemic unity of all humankind.

This relationship is found not only in the United States but also in various settings in modern times. Scientific universalism as a counterweight to ethno-racial prejudice appealed to some people in contemporary Europe, too. Indeed, the notion that science can bring everyone together is an old and powerful idea, comparable in a secular idiom, as I have suggested elsewhere, to the Christian story of Pentecost.[18] Yet, at the relevant historical moment, the United States was a distinctive setting. Unlike Germany or France or Britain or Poland or Italy, it was a heavily immigrant society and the process of immigration brought substantial numbers of Jews into a society long dominated by a dissenting Protestantism more intense than the prevailing religious regime in other industrialised nations of the North Atlantic West. The dynamics of anti-Semitism and of philo-Semitism were affected by these conditions. Anti-Semitism responded to sharp demographic changes, while philo-Semitism accelerated dialectically as Jews were recognised as convenient means for undermining the authority of an old Protestant cultural hegemony that many liberals and radicals found confining.

The distinction between the generic, enduring, scientific 'attitude', on the one hand, and, on the other, the specific contents of scientific knowledge at any given point in time turned out to be profoundly important. For many American scientists and scholars at the beginning of the interwar era, being 'scientific' meant accepting the apparent truths of eugenics in particular and obeying,

---

17 For examples of the New York office of Soviet intelligence reporting to Moscow on William Malisoff's activities as a secret agent, see two dispatches by Soviet intelligence agents that were declassified in the 1990s and are now known as the 'Venona Transcripts', after the code name for the American surveillance project, 'Venona', available on the Internet as Venona #1276 (2 August 1943) and #1680 (30 November 1944). Malisoff's extensive work for the KGB is summarised in John Earl Haynes and Harvey Klehr, *Venona: Decoding Soviet Espionage in America* (New Haven, 1999), 191.

18 David Hollinger, 'Science as a Weapon in *Kulturkämpfe* in the United States during and after World War II', *Isis* 86 (1995), 440; Hollinger, *Science, Jews, and Secular Culture* (n. 2), 156.

in general, the laws of nature as explicated by elite scientists. A portentous example of this style of advocacy of a scientific civilisation is Madison Grant's enormously popular book *The Passing of the Great Race* (1916), the fourth edition of which appeared in 1921 with an elaborate scholarly apparatus that better enabled it to be the most widely quoted work in the debate, culminating in the Immigration Restriction Act of 1924.[19]

Grant's white supremacist and anti-Semitic ideas were widely accepted as essentially sound by much more of the scientific community than historians of science have generally acknowledged. Jonathan Spiro has shown how eager leading biologists were to help Grant fix the mistakes in his early editions so that he would not be vulnerable to attack by technically sophisticated critics. Literally scores of scientists, including the chair of zoology of Harvard and of biology at Princeton, worked closely with Grant to make sure that this flagship for science would be strong enough to withstand the politically motivated efforts to sink it, which they all knew would come from Jews and their allegedly soft-headed defenders.[20] The scientists and scholars who supported Grant – like Henry Fairfield Osborn, head of the American Museum of Natural History, President C.C. Little of the University of Michigan and the Yale geographer Ellsworth Huntington – were all loquacious advocates of a more scientific civilisation in the United States. For them, this was not a methodological attitude of continuing inquiry. Rather, a scientific civilisation for them entailed a relation of social obedience to the hard and fast results of science so often ignored by sentimentalists.

The ranks of these obey-nature-style American prophets of a scientific future included some religiously unorthodox products of Protestant culture, including Grant himself, but these ranks included few Jews, few non-Nordic immigrants and few socialists. The most negative reviews of Grant's book, by contrast, were produced by two Jewish friends of Dewey's: Franz Boas and Horace Kallen.[21] The latter, interestingly, was to emerge some years later as a sharp critic of the *International Encyclopedia of Unified Science* as excessively totalitarian and as much too narrow in its own fashion as to what shall count as science.[22] But my point here is that the idea of pulling together all of the results of science and trying to organise society on a scientific basis was being put forth

---

19 Madison Grant, *The Passing of the Great Race* (New York, 1916).

20 Jonathan Peter Spiro, *Defending the Master Race: Conservation, Eugenics, and the Legacy of Madison Grant* (Burlington, 2009).

21 Horace Kallen and Franz Boas, reviews of Grant, *The Passing of the Great Race*, as cited in Spiro, *Defending the Master Race* (n. 20), 367–8.

22 Kallen's critique of the unified science movement as 'totalitarian' appeared in articles in 1940 and 1946. See George Reisch, *How the Cold War Transformed Philosophy of Science: To the Icy Slopes of Logic* (New York, 2005).

in the late 1910s and early and mid-1920s in the United States by a formidable cohort of largely nativist, Anglocentric intellectuals whose spirit was caught in a biologist's assertion in the pages of *Science* in 1924 that the commandments of eugenics are as valid as those of Sinai.[23]

Dewey's pragmatic emphasis on method and attitude was directed against the contrasting emphasis on the solidity of scientific knowledge, which Dewey believed tended to render in absolute terms a body of knowledge that was always changing. The racist intellectuals who treated contemporary eugenics as if it were written in stone displayed a view of scientific knowledge entirely alien to Dewey. Jewish followers of Dewey were highly sensitive to the distinction and glad to help undermine the claims to absolute truth offered on behalf of contemporary racist ideas. Kallen, a Jewish protégé of Dewey's who wrote some of the most important articles for the *Encyclopedia of Social Sciences*, fits this pattern exactly. Even more prominent in this contingent of science-preoccupied intellectuals of Anglo-Protestant or Jewish origin was Morris R. Cohen, the first Jew to obtain a philosophy PhD at Harvard, to become President of the Eastern Division of the American Philosophical Association and to be elected one of that organisation's Carus Lecturers. Cohen's 1923 edition of Charles Peirce's writings, *Chance, Love, and Logic*, established the core of the Peirce canon and presented Peirce as an exemplar of the scientific attitude and a prophet of 'liberal' civilisation. Cohen was commissioned to write the pivotal article for the *Encyclopedia of Social Sciences*, the one on 'Scientific Method', and was, along with his student Nagel, the co-author of *Introduction to Logic and Scientific Method*, the most widely used textbook for a quarter-century. I would say more about Cohen, whom the Vienna émigrés appreciated as the leading philosopher of science in the United States, despite his metaphysical commitment to logical realism, had I not already devoted an entire book to Cohen's engagement with science as a cultural ideal and to his influence on the succeeding generation of philosophers and scientists, many of whom he taught at the City College of New York and the University of Chicago, where Cohen shared an office with Carnap.[24]

Instead, I will focus on the gentile side of the Jewish-gentile left-liberal intelligentsia of which Cohen and Dewey were the leading philosophers during the interwar decades. I will take up two important artefacts, the novel *Arrowsmith* (1925), which was the chief basis on which Sinclair Lewis became

---

23 A.D. Little, 'The Fifth Estate', *Science* 60 (1924), 299–306, esp. 304.

24 For Morris Cohen's standing among the philosophers of the Vienna Circle, see Herbert Feigl, 'The Vienna Circle in America', *Perspectives in American History* 2 (1968), 660, and David A. Hollinger, *Morris R. Cohen and the Scientific Ideal* (Cambridge, MA, 1975). There is no record in the Morris R. Cohen Papers at the University of Chicago or elsewhere known to me of any effort on Rudolf Carnap's part to incorporate Cohen into the unified science endeavour.

the first American to win the Nobel Prize for Literature, and the 1919 article, 'The Intellectual Preeminence of Jews in Modern Europe', by Thorstein Veblen, the leading American social theorist of the generation. I choose these two examples because of the exceptional importance of these two authors to the culture of American intellectuals during the interwar decades and because both register resoundingly the capacity of scientific inquiry to bring together Jews and non-Jews. By examining these two texts in some detail, we can engage aspects of the American scene that Neurath and others of the Vienna Circle were slow to comprehend.

Arrowsmith is a perfectly dreadful novel, but it is a document of some importance. Its theme is the struggle of a young Midwesterner to find a calling commensurate with the moral and spiritual capabilities granted to him by nature and by his Protestant and pioneer heritage. Only in pure science is this capacity finally fulfilled. The young Martin Arrowsmith encounters science through contact with the great bacteriologist and immunologist Professor Max Gottlieb, a Jewish émigré from Germany.

As soon as he is left alone after his first, incandescent meeting with Gottlieb, Arrowsmith prays for the strength to live by the austere values for which his idol stands. In what Lewis calls the 'prayer of the scientist', Arrowsmith asks God to endow him with the energy and will to labour relentlessly within the strict rules of his calling and then asks for the strength to trust in God for nothing else.[25] This moment of dedication defines the rest of the narrative, during which Arrowsmith periodically weakens and is drawn away from the ideals of truly scientific behaviour, is haunted by his conscience and returns finally to renew his dedication and to act decisively upon it by rejecting all human society except for a single, like-minded co-worker with whom he retreats to a laboratory in the Vermont woods, seizing a life that Lewis depicts with images drawn from Arrowsmith's pioneer grandparents as they went west in a covered wagon.

This tremendously popular piece of fiction – voted in 1944 by readers of the *Saturday Review of Literature* to be the best American novel in the previous twenty years, beating out all the classics of William Faulkner, F. Scott Fitzgerald, Ernest Hemingway and John Dos Passos – invokes two of the most powerful myths in all of American history – the myth of the Protestant saint and the myth of the pioneer-frontiersman – and consolidates them in the figure of an immigrant German-Jewish scientist.[26] Lewis's novel renders the potentially alien ideology of pure science as American as Mark Twain. *Arrowsmith* recalls *The Adventures of Huckleberry Finn* and *The Deerslayer*, even to the point of accepting as a social ideal the exclusive and robust comradeship of two boyish males. In Lewis's closing frame, Martin and his lab partner are pictured like Huck

25 Sinclair Lewis, *Arrowsmith* (New York, 1925), 269.

26 'SRL Poll on Novels and Novelists', *Saturday Review of Literature*, 5 August 1944, 61.

and Jim, 'far out on the water' of a tree-lined lake in a 'clumsy boat', exulting in their freedom.[27]

The arrogance of the German scientific tradition embodied in Gottlieb could easily have threatened the no-nonsense social egalitarianism that Lewis was always at pains to affirm. But by making Gottlieb Jewish, Lewis gave himself a chance to depict Gottlieb as both a victim of the arrogance and prejudice of others and as a symbol for the tolerance and internationalism of science. Gottlieb's own arrogance, moreover, is carefully constructed: The great professor scorns only pretenders to intellectual stature. He develops an implausible friendship with the ostentatiously red-blooded, self-made American industrialist Ross McGurk, who enjoys sneaking away from his busybody wife in order to lunch alone with his fellow supermale, Gottlieb the scientist. And it is Gottlieb, the lofty, forbidding symbol for true *Wissenschaft*, who comes 'naturally' to Martin's mind when the youth finds himself in Montana, among electrical workers atop telephone poles on the prairie, who 'had no pretentiousness' and were 'as healthy and as simple as the west wind'.[28]

Moreover, Gottlieb's successor as Martin's alter ego is the indefatigably down-to-earth Terry Wickert, the 'red-headed, roughfaced, wiry' bachelor with whom Arrowsmith moves to Vermont.[29] Just when Wickert and Arrowsmith have apparently internalised Gottlieb's values to the point that the old Jew's moral tutelage is no longer essential, Gottlieb passes into senility. The two unambiguously American good old boys can now themselves exemplify the scientific ethic, with a little time out now and then for fishing and for watching the frolic of squirrels. Arrowsmith himself is depicted as an uncompromising social egalitarian and as a product of the ethnic melting pot. If the novel is not overtly democratic, the elitism it espouses is one that can be housed in an old farmhouse in the woods and one to which anyone with brains and will can reasonably aspire.

Will, more than brains, is what matters most. *Arrowsmith* is a novel more about character than about intelligence, more about the morality of the scientific enterprise than about its intellectual results. Indeed, at the end of the novel Lewis treats the actual results of research as much less important than organising one's life around the quest for sound results. Science was an agent of cultural integration not on account of the knowledge that citizens might learn, but on account of its ability to unite persons of various backgrounds into a single and noble discipline of the mind.

That Jews did this better than anyone else was affirmed by Veblen in 1919 in the essay that remained, until the 2004 appearance of Yuri Slezkine's *The Jewish Century*, the most widely discussed treatise ever written on the overrepresentation

---

27 Lewis, *Arrowsmith* (n. 25), 430.

28 Ibid., 33.

29 Ibid., 275.

of Jews among leading scientists and scholars. Veblen had no use for a genetic explanation for Jewish intellectual achievement, nor did he offer an explanation grounded in traditional Judaic respect for learning and Talmudic disputation. For Veblen, the key factor was the detachment he attributed to Jews as a people in between traditional Jewish communal life and modern gentile society; the combination of withdrawal from Jewish tradition and only partial assimilation into gentile society endows the intellectual Jew with the virtue of detachment. Sceptical, estranged and alienated, the liberated Jew, as Veblen described this distinctive character in the drama of modern European history, was ideally suited for a career of science and scholarship. This marginal man, this 'wanderer in the intellectual no-man's land', could see 'unmediated facts' that were less readily observed by minds still clouded by Judaism, by Christianity, by local customs, by tribal idols and by other pre-modern frames of reference.[30]

These emphases of Veblen's troubled some readers. Indeed, the very first person to read Veblen's essay was thoroughly appalled. This was the editor of a Zionist magazine in New York who had commissioned the piece after hearing Veblen hold forth over lunch about the extraordinary intellectual contributions of Jews. He rejected the article for publication, and that is why it ended up in the pages of the *Political Science Quarterly*. The editor who originally commissioned the piece seems to have expected, on the basis of the table-talk of this Wisconsin-born semitophile, a celebration of Jewish tradition, a proto-Zionist, race-oriented explanation in terms of Jewish blood and Jewish tribal solidarity. What this well-meaning Zionist editor got instead was what almost every Veblen scholar from Joseph Dorfman through Lewis Feuer and John P. Diggins to the present has agreed was Veblen's self-portrait. Veblen, the son of Norwegian immigrants, elevated the apotheosis of alienation into an art form, and nowhere did he practice this art more commandingly than when he projected his alienated self onto Jewish intellectuals in this remarkable essay.[31]

But the single most striking property of Veblen's text is its silence about the economic activities and economic position of Jews. This silence might be construed as all to the good insofar as it entails a refusal on Veblen's part to accommodate and retain anti-Semitic stereotypes. The era's greatest scourge of the pecuniary withheld his fire when he spoke about Jews. Veblen's essay

---

30 Thorstein Veblen, 'The Intellectual Preeminence of Jews in Modern Europe', as reprinted in Max Lerner (ed.), *The Portable Veblen* (New York, 1948), 475. This essay first appeared in *Political Science Quarterly* in 1919. The discourse about Jewish pre-eminence has been fundamentally changed by Yuri Slezkine, *The Jewish Century* (Princeton, 2004), which emphasised rather than avoided the commercial experience of Diaspora Jews.

31 For Thorstein Veblen's interaction with editors, see Lewis Feuer, 'Thorstein Veblen: The Metaphysics of the Interned Immigrant', *American Quarterly* 4 (1953), 106, and John P. Diggins, *The Bard of Savagery* (New York, 1978).

of 1919 is in no way a Jewish postscript to his slashing book of twenty years earlier, *The Theory of the Leisure Class*. The wealthy parasites of Veblen's most enduring work, those exemplars of conspicuous consumption who seduce the masses from virtuous industry, those predatory barbarians, are not Jewish. Rather, they are the friends and confidants of Madison Grant. Had Veblen been less hostile to commercial culture and less determined to employ Jewish intellectuals as counter-models to it, he might have explored the possible relevance of the class position of Jews to his inquiry.

His having stayed well away from any such explanation speaks all the more compellingly to the philo-Semitism of the essay. Veblen, who in that same year of 1919 collected his scattered papers under the title *The Place of Science in Modern Civilization*, had for years been extolling the very sensibility he now attributed more to Jews than to anyone else. Just when contemporaries were associating Jews with Bolshevism and with the insidious undercutting of the place of Nordics in the United States, Veblen urged gentile empathic identification with Jews and depicted Jews as the unique carriers of the scientific attitude.[32]

When Neurath came to New York, then, he did not have to go far to find a sprawling community of Americans going sort of his way, although he seems to have had difficulty connecting with this community, except as mediated by the highly sectarian project of logical empiricism. In New York and beyond was an imposing dispersion of philo-Semitic gentiles and immigrant Jews, democratic socialists and Stalinists and antifoundational pragmatists and experienced encyclopaedists, all united in their devotion to what they called the scientific attitude.

At the ecumenical centre of this dispersion of science-affirming intellectuals was Dewey himself, who opened his 1938 essay for Neurath's *Encyclopedia* with a description of the many areas of life in which that attitude could be found and with a call for unity. The call for unity was accompanied by the following plea: '[A] movement in behalf of the unity of science need not and should not lay down in advance a platform to be accepted'. To formulate the bases of unified science too narrowly, said Dewey to his milieu's most conspicuous platform-drafters, 'is both to obstruct cooperation and to be false to the scientific spirit'. Dewey concluded with a series of bland reminders that science was the great enemy of national and racial prejudice, that science was inherently universal and that its method rather than its results was the basis for this unity.[33] It was

---

32 I have addressed Veblen's legendary essay more extensively in David A. Hollinger, 'Why Are Jews Preeminent in Science and Scholarship? The Veblen Thesis Reconsidered', *Aleph* 2 (2002), 145–63, reprinted in Hollinger, *Cosmopolitanism and Solidarity: Studies in Ethnoracial, Religious, and Professional Affiliation in the United States* (Madison, 2006).

33 John Dewey, 'Unity of Science as a Social Problem', in Otto Neurath et al., *Encyclopedia and Unified Science* (Chicago, 1938), published as vol. 1, no. 1, of *International Encyclopedia of Unified Sciences*, 29–38, esp. 33–34.

old stuff for Dewey. Older than Neurath probably realised. Dewey slipped the no doubt unsuspecting Neurath an essay some of which was, almost word for word, what he had published in an article in *Science* more than a quarter-century before, in 1910.[34] That's about where, in the chronological development of ideas and programmes connected to science, Dewey understood the well-meaning Neurath to be.

Neurath's death in England in 1945 radically altered the leadership of the movement for unified science and ensured that the legacy of this movement in the United States would be all the more tightly connected to Carnap, Nagel and others associated with logical empiricism. Their story – the story of the triumph of their style of philosophy over Dewey's in the leadership of American PhD-producing departments of philosophy – has often been told and continues to attract the close attention of disciplinary historians.[35] But that story is a distinctly disciplinary one, having to do with philosophy, narrowly considered, not with the larger struggles over the direction of culture in which philosophers of the logical empiricist tradition proved much less eager to engage than had Dewey and his followers and, in his own fashion, had Neurath himself.

Yet, as the logical empiricist philosophers went their own way, many other American intellectuals perpetuated in a variety of contexts the multifaceted movement to move the public culture of the United States in directions that were said to be 'scientific' and thereby to render that culture more open to Jews and by implication to other non-Christians. Prominent in that movement during the 1940s and 1950s were various leading academics, including Merton, Hook, Harvard President James B. Conant, historian Richard Hofstadter and countless social scientists. I will not repeat here the account of that phase of the movement, which I have addressed in an essay entitled 'Science as a Weapon in *Kulturkämpfe* in the United States during and after World War II'.[36] But I do want to observe that during that period the *International Encyclopedia of Unified Science* became increasingly isolated from the concerns of the bulk of scientifically engaged intellectuals in the United States.

An index of the *Encyclopedia*'s status as an anachronism is the fact that its most influential article appeared within its covers almost by accident, paid almost no homage to the 'Unity of Science' doctrinal heritage, took severe issue with

---

34 John Dewey, 'Science as Subject-Matter and as Method', *Science*, 36 (28 January 1910). Compare especially Dewey's opening discussion in the 1938 text, in Neurath et al., *Encyclopedia and Unified Science* (n. 33), 29.

35 For a recent example, see Bruce Kuklick, *A History of Philosophy in America, 1720–2000* (Oxford, 2002). For a more contextualised counter-example, see Chapter 11, this volume.

36 Hollinger, 'Science as a Weapon' (n. 18). See also David A. Hollinger, 'Religion, Ethnicity, and Politics in American Philosophy: Reflections on McCumber's *Time in the Ditch*', *Philosophical Studies* 108 (2002), 173–81.

some of the favourite ideas of logical empiricists and was rather complacently aloof from the culturally reformist ambitions of Neurath and Morris as well as of Dewey. The work to which I refer has been associated with the *Encyclopedia* mostly in the minds of those tempted to exaggerate the *Encyclopedia*'s historical importance and eager to claim this famous treatise for their case. I refer, of course, to Thomas S. Kuhn's 1962 book-length article, *The Structure of Scientific Revolutions*. Commissioned by Neurath's successors when the *Encyclopedia* was already on its last legs, Kuhn's essay was long in coming and, by the time it appeared, had a very different complexion from what those who commissioned an essay on the dynamics of scientific change envisaged. Kuhn's book became the most widely cited single work in the reaction against the positivistic ideas associated with the Unity of Science Movement. No one did more than Kuhn to convince the world that the progress of science depended on highly specific, historically embedded communities of inquirers, not on 'scientific method' of any logical structure. Hence, the only contribution to the *International Encyclopedia of Unified Science* to be cited with any regularity from the early 1960s onward was the one contribution that was the most subversive of its ruling ideas.

By the time Kuhn's work appeared, the ideal of science as a universal mode, excluding no one on account of religion or ethnicity, had done most of the work in the United States that the generation of Cohen and Veblen and Lewis and Dewey had in mind for it. The epistemic universalism of science had helped to diversify the community of knowers and placed on the defensive those who resisted this diversification. Among the achievements of this universalism was facilitating the entry into American academic leadership of those philosophers of the Vienna Circle who fled Hitler's Europe and creating an atmosphere in which people like Kuhn, whose status as a Jew would have been an issue a generation before, could be at the centre of the national academic establishment and could even become the foremost theorist of how that establishment's scientific communities actually worked. In the long run, the Unity of Science Movement as led by Neurath and his closest colleagues had its greatest American impact on one disciplinary community, that of philosophy, but the larger movement for an American culture organised around the universalist ideals of science, of which the Unity of Science endeavour was but one particular and decidedly sectarian segment, made American academia as a whole a much different place than it had been when Dewey and his contemporaries began their labours.

Acknowledgement

This chapter is reprinted from *Pacific Historical Review* 80 (2011), 211–230, with kind permission of the University of California Press. For critical suggestions, I wish to thank Bruce Kuklick and George Reisch. For assistance with documentation, I wish to thank Noah Strote.

Chapter 11

McCarthyism in philosophy and the end of the unity of science ideology

George Reisch

There is a popular myth that the scientific philosophy known as logical empiricism emigrated to America without the socialist political and cultural agenda it promoted in Austria and Germany.[1] In fact, however, the new philosophy of science came to North America hand-in-hand with Otto Neurath's Unity of Science Movement – a movement that encompassed much more than a philosophical endeavour: together with its stance against metaphysics and for a strictly empiricist epistemology, the movement promoted a scientific conception of the world with the aim of using unified science as a tool for social planning.[2] Its project was collectivist, internationalist and socialist in intent.

This agenda was not dropped as soon as the Unity of Science Movement crossed the Atlantic, after most of the members and associates of the Vienna Circle had fled central Europe in the 1930s upon the rise of National Socialism. The political dimensions of the project began to disappear several years later as a result of a transformation that occurred on American soil and for distinctly American reasons. Here I examine this transition by contrasting American logical empiricism and American philosophy in the mid to late 1930s with logical empiricism, Cold-War style, of the 1950s and 1960s. I aim to show that the difference is striking and profound. I shall also offer some evidence that the reasons for this transformation have much to do with the politics – international, intellectual and personal – of the Cold War.[3]

---

1 For example, R. Giere, 'From *Wissenschaftliche Philosophie* to Philosophy of Science' in R. Giere and A. Richardson (eds), *Origins of Logical Empiricism* (Minneapolis, 1996), 335–54: 338; P. Galison, 'Constructing Modernism: The Cultural Location of Aufbau', ibid., 17–44: 40. Logical empiricism is also commonly known as logical positivism.

2 I have argued this point in detail in G. Reisch, 'Planning Science: Otto Neurath and the *International Encyclopedia of Unified Science*', *British Journal for the History of Science* 27 (1994), 153–76.

3 For a more detailed presentation of the claims made in this paper, see G. Reisch, *How the Cold War Transformed Philosophy of Science: To the Icy Slopes of Logic* (Cambridge, 2005).

Logical empiricism and the unity of science movement in America

When the Unity of Science Movement arrived in New York, Chicago and Boston in the 1930s, it was very much like its original European version. As an organisation, it in effect replaced the Ernst Mach Society, through which the Vienna Circle had sponsored public lectures before the dispersal of its members. Led by Otto Neurath (living in Holland and, from May 1940, England), Rudolf Carnap and the American pragmatist Charles Morris (both at the University of Chicago), the movement consisted in annual congresses and publications including the *International Encyclopedia of Unified Science*. All these activities were aimed at promoting and popularising the enlightenment agenda that Neurath, Carnap and Hans Hahn had outlined in the Vienna Circle's manifesto, *Wissenschaftliche Weltauffassung*.[4] In America, Neurath and his movement made their appearance twice each in both *Time* and the *New York Times*. The articles in the latter were written by Neurath's cousin Waldemar Kaempffert, who was one of the leading science writers in the country.[5]

One reason the Unity of Science Movement thrived in 1930s America was that most intellectuals at the time were socialist. On the heels of the Great Depression, many believed that capitalism and free-market economics were fundamentally flawed and saw the Soviet Union as a bold, promising experiment in socialism and scientific social planning, especially at the time of Lenin's leadership. Agreeing that comprehensive scientific planning required a comprehensive, or unified, science, most of these intellectuals automatically respected Neurath and the Unity of Science Movement as ideological allies.

When Neurath visited New York in the autumn of 1936, Ernest Nagel acted as his main host. After one social gathering – quite possibly a reception for Neurath himself – that was attended by most of the New York philosophers, Neurath asked Nagel to write up a short inventory of those who were there. Along with his descriptions of their jobs and academic specialities, Nagel described these philosophers' politics: who was a devotee of dialectical materialism, who was more positivistic and so on. Most, Nagel wrote, 'were Jewish and I think without exception have left sympathies in politics'.[6]

Neurath's trip was a success. He recruited some of these philosophers, including John Dewey, to write for the *International Encyclopedia of Unified Science*, and they all seemed to like Neurath very much. Neurath hit it off with

4 Verein Ernst Mach, *Wissenschaftliche Weltauffassung: der Wiener Kreis* (Vienna, 1929).

5 Kaempffert's articles appeared in the *New York Times*, 14 February 1937, and the *New York Times Book Review*, 7 April 1938. See also unsigned articles in *Time*, 1 August 1938 and 18 September 1939.

6 Nagel to Neurath, 13 October 1936, Otto Neurath Nachlass (hereafter: ONN), Rijksarchief, Noord-Holland, Haarlem, the Netherlands.

William Malisoff, the founder and editor of *Philosophy of Science*, who published articles and announcements about the Unity of Science Movement in that journal.[7] On the other hand, or more precisely, what would *become* the other hand, Neurath also got along well with Dewey's student Sidney Hook. Some two and a half years later, Hook sent Neurath – then living in Holland – a short letter in which he wrote 'we talk about you here in New York often and regard you as one very close to us indeed'.[8]

This 'we' who talked about Neurath was probably the extended circle of New York intellectuals, comprising not only philosophers but also writers and critics such as Lionel Trilling, Edmund White, Granville Hicks, James T. Farrell and John Dos Passos. Besides some correspondence between Farrell and Neurath, evidence for this is also found in the Marxist journal *Partisan Review*, a popular forum for the New York intellectuals. In 1939, two articles praised logical empiricism as an ally in the struggle for a more socialistic, rational and literary world.[9] In the winter issue of that year William Gruen, who had become familiar with logical empiricism when taking Carnap's seminar at Harvard in the summer of 1936, wrote:

> Marxism as an empirical social *science* is also Marxism as a *method* of revolutionary action.... This is the practical bearing of the logical empiricist analysis of meaning. The need for social action demands that social science be unencumbered by doctrines which have no active, operative significance. To this end logical empiricism offers a powerful critique which has repeatedly proved its effectiveness in the analysis of natural science, but whose full advantages still await realization in the field of esthetics, ethics, and political thought.[10]

In the previous issue, Albert Wohlstetter and Morton White praised the Unity of Science Movement for its serious study of semantics, its anti-metaphysical stance and its aim of the integration of science 'to be illustrated in the collective project of many scientists, the Encyclopedia of Unified Science'.[11]

Logical empiricism also made a splash in the first issue of another journal, *Science & Society*, which embraced the scientism of the Unity of Science Movement and its leftist politics but also went further in its explicit Marxist orientation. If *Partisan Review* aimed to build a modern Marxist, literary and

---

7 See also Chapter 10, this volume.

8 Hook to Neurath, 27 June 1939, ONN.

9 A. Wohlstetter and M.G. White, 'Who Are the Friends of Semantics?', *Partisan Review* 6 (1939), 50–57, and W. Gruen, 'What is Logical Empiricism?', *Partisan Review* 6 (1939), 64–77.

10 Gruen, 'What is Logical Empiricism?' (n. 9), 77.

11 Wohlstetter and White, 'Who are the Friends of Semantics?' (n. 9).

scientific worldview, *Science & Society* confined itself to Marxism, social science and the recruiting of intellectual talent for the coming revolution. As its inaugural editorial announced, the journal would 'solicit the active cooperation of all who are interested in its [Marxism's] progress' and to encourage those who are vacillating or unsure about 'the necessary connection between thought and action' to get off the fence. They would 'find the answer to their doubts through their participation in this journal'.[12]

There were more than just conceptual and ideological connections between *Science & Society* and Neurath's Unity of Science Movement. In its first issue V.J. McGill, a young professor of philosophy at Hunter College, presented '[a]n evaluation of logical empiricism' from an explicitly dialectical-materialist perspective.[13] McGill followed Lenin's critique of empiricism and complained, in short, that logical empiricism was incorrect precisely to the extent that it was not materialist. Shortly after McGill's article appeared, however, he met Neurath at the reception mentioned earlier, and it seems likely that Neurath's famous personal charm may have had something to do with McGill's increasingly sympathetic attitude toward the movement. Three issues later, for example, in an article entitled 'Logical Positivism and Unity of Science', he noted that 'Marxists are also committed to the unification of science' and called for cooperation between the two groups on the grounds that Carnap's Principle of Tolerance 'should be extended to the realist idiom'.[14] Then there would be 'special advantage in admitting realists and Marxian materialists to the discussions of the Congress for the Unity of Science'.[15]

Several other features show just how politicised this intellectual culture was. *Science & Society* reached out to Soviet intellectual culture by appointing as one of its editors the American geneticist Hermann J. Muller, then living and working in Leningrad. Besides American intellectuals, the roster of editors on the journal's masthead included well-known European radicals: French physicist Paul Langevin and, from Britain, J.D. Bernal, Lancelot Hogben, Hyman Levy and Joseph Needham. Among its regular contributors was Dutch émigré and ardent Marxist Dirk Jan Struik. The journal itself was initially edited, and probably co-founded, by Albert Blumberg, teacher of philosophy at Johns Hopkins University. Blumberg was born in America, finished his graduate work under Moritz Schlick in Vienna, and returned to the United States to introduce many Americans to scientific philosophy by writing, together with

---

12 Unsigned editorial, *Science and Society* 1 (1936), i–ii.

13 V.J. McGill, 'An Evaluation of Logical Empiricism', *Science & Society* 1 (1936), 45–80.

14 V.J. McGill, 'Logical Positivism and the Unity of Science', *Science & Society* 1 (1936), 550–61.

15 Ibid., 561.

Herbert Feigl, the article 'Logical Positivism: A New Movement in European Philosophy'.[16] By 1937, Blumberg had dropped out of academic philosophy and become a rapidly rising star in the American Communist Party. From the late 1930s through the mid-1950s, he and his wife, Dorothy Rose, worked at the highest levels of power in the party, usually in close proximity to and routinely in contact with leaders such as William Foster and Earl Browder. Both Blumbergs were convicted of treason in 1956, although Albert did not serve time because his conviction was overturned when the Smith Act, under which he was found guilty of treason, was itself judged to be unconstitutional.

Blumberg's counterpart at *Philosophy of Science*, William Malisoff, took a slightly different and earlier route to socialism and communism. Until his premature death in 1947, Malisoff remained a chemist, author and editor who also happened to be a spy for the Kremlin. He was investigated by the Federal Bureau of Investigation (FBI). His file and some of the Venona documents (decoding Soviet communications during the Second World War) suggest that Malisoff was of interest to Moscow for technical chemical information, possibly related to atomic research, that he might provide to them.[17]

In summary, many American philosophers before the Cold War were as politically engaged, as left-wing and as interested in science as a means of understanding and changing the world as were the founders of the Unity of Science Movement.[18] Hence, they were very much open to all dimensions of the project that Neurath and his colleagues brought to America.

Apolitical philosophy of science

Twenty years later, at the end of the 1950s, American philosophy of science was quite a different business. The decidedly leftist intellectual culture of the 1930s, ranging from Blumberg's and Malisoff's communist activism to the more moderate 'left sympathies' that Nagel attributed to 'most' of the philosophers who received Neurath, gave way to an avowedly apolitical, professional milieu in which politics was treated at most as a matter of private opinion. Politics had nothing to do with philosophy of science, and the Unity of Science Movement was effectively dead.

16 A. Blumberg and H. Feigl, 'Logical Positivism: A New Movement in European philosophy', *Journal of Philosophy* 28 (1931), 282–96. This information about Blumberg's political career comes from his FBI file, a copy of which was obtained by the author under the Freedom of Information Act.

17 J.E. Haynes and H. Klehr, *Venona: Decoding Soviet Espionage in America* (New Haven, 1999), 290–91.

18 See also Chapter 10, this volume.

One indication of this transition is that, as shown by Don Howard's analysis, the journal *Philosophy of Science* featured fewer and fewer articles addressing matters of politics, society, or ethics.[19] Other indications are the testimonies of two leaders in the profession, both of whom explicitly demarcated philosophy from politics. Herbert Feigl, whose Minnesota Center for the Philosophy of Science has been one of the main institutional centres of the profession since the 1950s, made this point in his introduction to the profession-defining book *Current Issues in the Philosophy of Science*.[20] One issue that was not included, Feigl emphasised, was that of 'our fundamental obligations or . . . the supreme goals of life.' While acknowledging that science can of course empirically describe social or individual goals and aspirations, he continued:

> But it must not be thought that such description and interpretation . . . could possibly establish their moral justification. Scientific truth is ethically neutral. . . . Once we have decided what we ought to do, science may be able to tell us what means will most effectively (and with the least interference with other morally authenticated purposes) enable us to do it. But there is no way of deducing moral imperatives from the truths of science.[21]

Most likely with a nod to those who wished the profession had retained some measure of engagement with social and moral issues, Feigl added that this situation was a given and that these professional, disciplinary limitations 'cannot be made into a charge against the scientific enterprise nor against a genuinely scientific outlook in philosophy. This would be like reproaching a weaving loom for its incapacity to produce music'.[22]

The editor of *Philosophy of Science*, Richard Rudner, said much the same thing some years later. In his book *Philosophy of Social Science* in the Prentice-Hall series of philosophy textbooks, Rudner began his presentation by distinguishing philosophy of social science from social philosophy:

> Where the social philosopher is concerned with the *tenability* of some one or another theory of social phenomena, the philosopher of science is concerned with the scientific *testability* of such theories. Testability is a necessary but not a sufficient condition of tenability. To go beyond concern with testability in the direction of ascertaining the tenability of a theory of social phenomena would be

19 D. Howard, 'Two Left Turns Make a Right: On the Curious Political Career of North American Philosophy of Science at Midcentury', in G. Hardcastle and A. Richardson (eds), *Logical Empiricism in North America* (Minneapolis, 2003), 25–93.

20 H. Feigl and G. Maxwell (eds), *Current Issues in the Philosophy of Science* (New York, 1959).

21 Ibid., 16.

22 Ibid., 16.

> for the philosopher of social science to enter the domain of the social scientist or the social philosopher.[23]

Philosophy of social science, in other words, was the closest professional philosophy of science could get to social philosophy, but it still did not get there. Philosophy of science was about structural, semantic and pragmatic features of scientific theories, which in turn had nothing to do with debates or theories concerning 'the worth of some social arrangements'.[24]

There was, moreover, at least one public relations manager who in the late 1950s believed that logical empiricism was so far removed from its original leftist socialism that he or she upheld one prominent logical empiricist, Hans Reichenbach, as an emblem of Cold War military research. In a series of advertisements appearing in *Science*, one can find stylised, full-page portraits of intellectuals as diverse as Lucretius, Leonardo, Francis Bacon, William Whewell and Henri Bergson. These are advertisements for the RAND Corporation, which was spun off from the Douglass Aircraft Corporation with US Navy money in order to create a less overtly military environment for intellectual researchers working in military science. The ads suggest that neither scientists nor philosophers would compromise their reputation or historical promise were they to engage in military research. That bid was made to devotees of logical empiricism, too: one of the advertisements portrays Reichenbach and includes a quotation from his book *Atom und Kosmos* of 1930 (Figure 11.1).[25] Once abstracted from its original political agenda, it would appear, logical empiricism was put to new use here serving the interests of the military-industrial complex.

On the battlefront

The transition from 1930s-style leftist philosophy of science to 1950s-style 'apolitical' philosophy of science was driven by three kinds of pressures that philosophers, and indeed all American intellectuals, faced in the late 1940s and 1950s.

First, there were general intellectual pressures and fashions against 'collectivism' in economic and social theory, as evidenced, for instance, by Friedrich Hayek's influential and popular *Road to Serfdom*.[26] In the philosophy of science, the trend was illustrated by Karl Popper's writings against 'closed societies' and Marxist historicism and by the New York philosopher Horace Kallen, who attacked Neurath and the Unity of Science Movement as promoting

---

23 R. Rudner, *Philosophy of Social Science* (Englewood Cliffs, 1966), 3; emphasis in original.

24 Ibid., 2.

25 See, e.g., Advertisement, *Science*, 8 May 1959.

26 F.A. Hayek, *Road to Serfdom* (London, 1946).

Figure 11.1 Hans Reichenbach depicted in an advertisement for the RAND Corporation in *Science* magazine, 1959.

a kind of 'totalitarian' philosophy and methodology of science.[27] The political resonance of Kallen's argument is obvious.

---

[27] K. Popper, *The Open Society and Its Enemies* (London, 1945); idem, *The Poverty of Historicism* (London, 1957). H. Kallen, 'The Meanings of "Unity" among the Sciences', *Educational Administration and Supervision* 26 (1940), 81–97; idem, 'The Meaning of

Second were institutional pressures stemming from anti-communist investigations at most major colleges and universities in America beginning in the late 1940s and continuing through the 1950s and 1960s. According to John McCumber, around 30 philosophers were interrogated by federal, state and university officials.[28] Many more prominent scientists, such as Robert Oppenheimer and Edward Condon, were publicly scrutinised and ostracised for actual or suspected past membership of the Communist Party. These pressures affected the philosophy of science at least in so far as several of its leaders were investigated by J. Edgar Hoover's Federal Bureau of Investigation. The FBI investigated not only Blumberg and Malisoff, for obvious reasons, but also Carnap and Philipp Frank, at a time when both were in the letterhead stationary of the short-lived Institute for the Unity of Science, which Frank and Morris had set up in Boston to continue the Unity of Science Movement in the wake of Neurath's death in 1945. Acting on a rumour that Frank had come to the United States to organise communist politics (and not to lecture on physics and philosophy, which would have been, on this hypothesis, his official cover), Hoover's agents began investigating Frank in the summer of 1952. During an investigation that lasted roughly two years, agents interviewed Frank's colleagues and friends, examined his immigration and employment records, and called in their secret informants in Boston-area communist groups to assess the truth of this rumour. In the end, the FBI concluded that Frank was indeed just a physicist and philosopher, not a subversive ideologue. Along the way, however, Hoover's agents obtained the name of Rudolf Carnap as one who could shed light on Frank's European past. Before interviewing Carnap about Frank, they noticed Carnap's many endorsements of progressive and leftist causes in the pages of the international communist newspaper *The Daily Worker*. Carnap himself then became subject to investigation. In 1954 and 1955, agents across the country investigated him and his activities in Chicago, Princeton (where Carnap worked at the Institute for Advanced Study) and the University of California at Los Angeles (where he took an appointment in 1954). As in Frank's case, agents concluded that Carnap was not a threat to American democracy.[29] But these exonerations came only after friends, neighbours and fellow philosophers of science were interviewed by Hoover's agents. If these interviews put in question Frank's and Carnap's politics and patriotism, one point was not in question to those associates and philosophers who found FBI agents ringing their

---

"Unity" among the Sciences, Once More', *Philosophy and Phenomenological Research* 6 (1946), 493–6.

28 J. McCumber, 'Time in the Ditch: American Philosophy and the McCarthy Era', *Diacritics* 27 (1996), 33–49.

29 This information about the investigations of Frank and Carnap comes from the respective FBI files, obtained by the author under a Freedom of Information Act request.

doorbells and asking questions about Frank and Carnap: rumour or the mere appearance of sympathy with communist politics could invite official scrutiny and investigation from federal, state or local investigators.

These pressures created a 'climate of fear' on most campuses, in which – as University of Chicago President Robert Hutchins put it in the popular magazine *Look* – teachers were 'afraid to teach'. Professors avoided discussing in classrooms anything possibly 'controversial', for once someone raised a red flag, so to speak, about a teacher, they were usually ostracised professionally and socially, if not formally investigated (see Figure 11.2.).[30] Hutchins's point introduces the third kind of pressure on philosophers of science: personal pressures coming directly from colleagues who sought either to expunge politics from philosophy, or to ensure that philosophers were sympathetic with a specific and right kind of politics.

Hutchins' comment was a response to Sidney Hook, a figure who epitomises this third kind of pressure. Hook was a typical American intellectual in so far as he was once an avid socialist – in fact a Trotskyist – who became so disillusioned with Stalin and the Kremlin that, by the early 1940s, he had become an avid anti-communist. Many other intellectuals took the same path from extreme left in the 1930s to an extreme, anti-Stalinist right in the 1940s – among others the editors of *Partisan Review*, Arthur Schlesinger, the writers Arthur Koestler and Ignazio Silone, James Burnham and Communist Party operatives such as Whittaker Chambers.

Hook was arguably the most ardent anti-communist. Personally and in print he attacked intellectuals who did not take the anti-Stalinist path, and he aggressively supported the purges of teachers and the main rationale behind them, namely that teachers who did not actively renounce past or present associations with communism were derelict in their duty to enquire and teach in the service of truth. For these reasons and others – including Hutchins's long-standing reservations about the emphasis on science in the educational theories and reforms proposed by Hook's teacher, Dewey – there was no love lost between Hutchins and Hook. This is clearest in Hutchins' article when he paints Hook as an apologist for, and agent of, McCarthyite purges of teachers:

> Whittaker Chambers and Prof. Sidney Hook of New York University, both of whom proclaim themselves devotees of academic freedom, say, 'Don't worry; only a few teachers have been fired.' What has this got to do with it? The question is not how many teachers have been fired, but how many think they might be, and for what reasons. It is even worse than that: Teachers are ... afraid of getting into trouble, with resultant damage to their professional prospects and their standing in their communities. You don't have to fire many teachers to intimidate them all.[31]

30 R.M. Hutchins, 'Are Our Teachers Afraid to Teach?', *Look* 18 (1954), 27–9.

31 Ibid., 28.

LOOK

VOLUME 18, NUMBER 5 • MARCH 9, 1954

the University of Chicago, where Prof. James Cate was photographed at the time Hutchins was its head, a probe, the author recalls, left "paralyzing" effects.

ARE OUR TEACHERS **Afraid to teach?**

BY ROBERT M. HUTCHINS

Dr. Hutchins, former president of the University of Chicago, is one of America's most brilliant, independent and provocative educators. His many educational innovations have included the now famous Great Books course and the abolition of course credits. Today, a quarter century after he made his bow at Chicago at the age of 30, he is an associate director of the Ford Foundation.

EDUCATION is impossible in many parts of the United States today because free inquiry and free discussion are impossible. In these communities, the teacher of economics, history or political science cannot teach. Even the teacher of literature must be careful. Didn't a member of Indiana's Textbook Commission call Robin Hood subversive?

The National Education Association studied no less than 522 school systems, covering every section of the United States, and came to the conclusion that American teachers today are reluctant to consider "controversial issues." But what does that mean? An issue is a point on which the parties take different positions. A noncontroversial issue, therefore, is as impossible as a round square. All issues are controversial; if they were not, they would not be issues.

In Los Angeles, Houston and Pawtucket, a teacher would hesitate to mention UNESCO, the United Nations Educational, Scientific and Cultural Organization, because school authorities have made it plain that they are afraid of it. Since those who oppose UNESCO generally oppose the United Nations, the teacher should probably not refer to the U.N. either. Since those who oppose the U.N. believe that the United States should somehow isolate itself from world affairs, the teacher would be unwise to say very much about international relations. How, then, can teachers say anything worth saying about the world in which we live?

continued

Produced by LEO ROSTEN

Figure 11.2 One of Robert Maynard Hutchins's popular defences of academic freedom against the anti-communist 'climate of fear'.

Source: R.M. Hutchins, 'Are Our Teachers Afraid to Teach?', *Look* 18 (1954), 27–9.

The wrath of Sidney Hook

It is difficult to say which of these three kinds of pressure was most effective in depoliticising the philosophy of science. It is clear, however, that the wrath of Hook was unpleasant to experience. Two examples show that anyone in Carnap's, Morris's, or Frank's shoes in the 1950s would need to be circumspect and cautious about pursuing the Unity of Science Movement whose values and goals were sympathetic to, and in some ways embraced by, socialists and communists. Hook was a kind of anti-communist watchdog with a loud bark and painful bite.

One example returns us to *Philosophy of Science* under William Malisoff. In 1947, among Malisoff's many short book reviews, one simply glowed. It concerned the book *Soviet Philosophy: A Study of Theory and Practice* by Hunter College philosopher John Somerville.[32] 'Hurry, get this book!' It is 'a "must"', Malisoff enthused, because it showed what was 'essential in the midst of the almost insane distortions this subject of Soviet Philosophy has received from the professional enemies of the USSR' and because the book analysed and clarified 'the miserable term "totalitarianism"' that came to dominate antisocialist and anti-communist discourse during the Cold War.[33] That 'miserable term', incidentally, was the one Horace Kallen repeatedly used to criticise logical empiricism and the Unity of Science Movement.

While Malisoff, on the far left, adored Somerville's book, Sidney Hook, on the far right, hated it. For Hook, *Soviet Philosophy* was hardly more than 'soviet propaganda'. In a review entitled 'Philosophy and the police' in *The Nation*, he skewered Somerville's book as an apology for 'the party character of philosophy and science' in the Soviet Union, and skewered Somerville himself as a substandard intellect whose love for all things Soviet led him into logical and factual errors.[34]

Most telling, I suggest, is not the substance of this debate between Somerville and Hook but the manner in which it unfolded. Somerville responded in a letter to the editor explaining that his book sought neither to apologise for nor condemn Soviet philosophy.[35] His goal, rather, was to make Soviet philosophy and its practice understandable to Western eyes, to assist international intellectual understanding and, ultimately, to help avoid military confrontation. 'It is not necessary to agree with the Soviet system', he explained in an article related to his book, 'but it is very necessary to cooperate with it, to meet it half way in a friendly spirit, just as in the case of the British or the Chinese'.[36]

---

32 J. Somerville, *Soviet Philosophy: A Study of Theory and Practice* (New York, 1946).

33 W. M. Malisoff, review of *Soviet Philosophy: A Study of Theory and Practice* by J. Somerville, *Philosophy of Science* 14 (1947), 172.

34 S. Hook, 'Philosophy and the Police', *The Nation*, 15 February 1947, 188–9.

35 Somerville's letter to the editor and Hook's response appeared in *The Nation*, 19 May 1947, 555–6.

36 J. Somerville, 'Soviet Science and Dialectical Materialism', *Philosophy of Science* 12 (1945), 23–9: 29.

Yet Somerville's letter, without his knowledge or approval, was edited and in part substantively reworded. Even worse, *The Nation*'s editor, Frida Kirchwey, allowed Hook to respond to Somerville's edited letter and hammer his indictment yet a second time: Somerville's book was 'a shabby piece of apologetics' full of 'deceptions peddled by fellow-travelers of the Communist Party line'.[37]

Somerville, in turn, was not allowed to respond, and he was understandably furious. He turned to Malisoff, who printed his account of the attack and some reflections on its significance in *Philosophy of Science*. Although he wrote roughly two years before the main wave of Cold War anti-communist investigations began to sweep through American colleges and universities, Somerville anticipated one of the most unphilosophical features of what came to be called the Cold War's anti-communist 'hysteria':

> The real insult [of Hook's and *The Nation*'s attack on Somerville] is that offered to the public, which must sooner or later wake up to the fact, if only the hard way, as I did, that the standards still assumed to be present have in reality vanished. If we cannot take elementary honesty for granted, what can we take for granted when we try to apply the results of serious professional research to a problem of the gravest public concern?[38]

The 'elementary honesty' of substantive debate and exchange of considered opinions was disappearing. In the nascent climate of the Cold War, differences of intellectual opinion about Soviet culture and politics were being construed as, and thus transforming into, differences of national loyalty. As Hutchins emphasised several years later, this was a central feature in a precarious atmosphere in which teachers were 'afraid to teach'.

This was hard-ball politics in philosophy. If Hook did not personally edit Somerville's letter to make him look more like a Soviet fellow-traveller than he really was, he almost certainly had something to do with it. For Hook's goal – ironic though it may be for one who so championed freedom over totalitarianism – was either to persuade his opponents to abandon their leftism and help promote anti-communism, or to make them simply shut up and cease to engage in political discourse. That was Hook's agenda with Rudolf Carnap, to whom he wrote a letter in 1949 (the underlinings are Carnap's):

---

37 S. Hook, response to Somerville (n. 35), 556.

38 J. Somerville, 'Ethics and Social Science: Case History of a Sharp Practice', *Philosophy of Science* 14 (1947), 345–7: 347.

March 20, 1949

Dear Carnap,

I have just learned from Dr. Jacobs of the Voice of America that your name has been cited (presumably abroad) as a sponsor of the forthcoming Cultural and Scientific Conference for World Peace. I hope this is not true. If you actually have enrolled yourself as a sponsor, I am confident that you are unaware of the real auspices of the Conference. It is being run by people whose first act, if they came to power, would be to liquidate you and people like you. The Conference is as much a Communist Party Front as was the Congress at Breslau-Wroclaw last summer.

I am enclosing some materials which I earnestly beg you to read very carefully. My experience with this Conference is decisive.

I can very well understand how honest people can disagree with American foreign policy now as in 1940. But what would you have thought of anyone who in 1940 lent his name to the German-American Bund who were then agitating for peace? People will draw the same inference about you if you are a sponsor of the Waldorf Conference and remain a sponsor. If you want to agitate for peace there is no need to associate yourself with agents of a foreign power: you can do it independently.

This business is no ordinary thing, as you will learn by developments in the next few days. Anybody who is still a sponsor by the time the Party-line begins to sound off at the conference, will be marked for life as a captive or fellow-traveler of the Communist Party. You are not a Communist and you are not a political innocent.

Forgive me for writing so strongly. It is a measure of the respect I have for you and your life-work. There are mistakes and mistakes. This one can be positively fatal if not rectified.

Sincerely Yours,

Sidney Hook[39]

Indeed Hook made news in 'the next few days' as he orchestrated a counter-conference to protest this conference, held at New York City's famous Waldorf Astoria Hotel.[40]

At a time when universities were beginning to investigate and dismiss (alleged) Communist professors, when Hook was a prominent and public

39 E. Shapiro (ed.), *Letters of Sidney Hook: Democracy, Communism, and the Cold War* (New York, 1995), 128–9. Carnap's underlinings are indicated in his copy of the letter, ASP RC 088–38–10, Rudolf Carnap Collection (hereafter 'Carnap Collection'), Archive of Scientific Philosophy, University of Pittsburgh.

40 See, for example, 'Soviet Is Attacked at Counter Rally', *New York Times*, 27 March 1949.

champion of these purges[41], Hook was plainly warning Carnap that his career was at risk. As to his choice of the words 'liquidate' and 'fatal', Hook was not being hyperbolic. Stalin was no friend of freethinking intellectuals, as Carnap surely knew, yet the fact that Carnap lived in America did not eliminate potentially 'fatal' consequences of his political dissent. At the time Hook wrote his letter, anti-communist legislation was working its way through the United States Congress. The end result, the Internal Security Act (or McCarran Act) of 1950, contained provisions for detaining or arresting individuals deemed likely to engage in political subversion. If, as many believed, the Third World War was imminent, and if Carnap was widely regarded as an ally of Communists, then more than his career could be at risk.

In the end, Carnap refused to be bullied. He replied that the statement the conference's organisers had asked him to sign called for peace and dialogue between the superpowers, and that is just what he wished to endorse. Standing up to Hook and his efforts to orchestrate others against Carnap was not easy, however. Some 20 years later, Carnap recalled that, despite pressures from his 'friends', he was proud not to have withdrawn his name as sponsor of the conference.[42]

Conclusion

The main point of this chapter is that the Cold War helped to transform American logical empiricism from a socially and culturally engaged project into a socially and politically sterilised enterprise. More particularly, I have presented evidence that the dynamics driving that transition were in many cases powerful and personal. Philosophers of science did not, out of curiosity or desire for professional advancement, put their fingers in the air, assess political winds, and then choose to scale back their political views. In these and most likely other cases, they were actively intimidated and bullied, if not by Sidney Hook or the FBI, then by a 'climate of fear' in which one word from a student or colleague suggesting that one's politics were unsavoury could snowball into a life- and career-changing investigation.[43]

The result of this process has been the transformation of the philosophical and political project of logical empiricism and the Unity of Science Movement

---

[41] See, for example, 'Should Communists Be Permitted to Teach?', *New York Times*, 27 February 1949.

[42] Carnap recalled the incident when writing to Cedric Belfrage in the late 1960s. Carnap to Belfrage, 23 November 1968, ASP RC 027–23–51, Carnap Collection.

[43] For an extended survey of intellectuals affected by the anti-communism of the Cold War (and earlier), see E. Schrecker, *No Ivory Tower: McCarthyism and the Universities* (New York, 1986).

into the apolitical academic discipline that philosophy of science became in the 1950s and beyond, especially in Anglo-Saxon countries. Embedded in this discipline, logical empiricism lived on as a strictly internalist endeavour, as witnessed by leading textbooks such as Nagel's *The Structure of Science.*[44] Yet the exclusion of the unity of science ideology from the discipline was driven by another ideology, that of the Cold War. In terms of its range of practices, philosophy of science has been shaped by this legacy.

Acknowledgements

This work was supported by a grant from the National Science Foundation; grant number NSF SES-0002222. A more detailed account of this cold-war transformation is presented in Reisch, *How the Cold War Transformed Philosophy of Science*, (Cambridge, 2005).

---

[44] E. Nagel, *The Structure of Science* (New York, 1961).

Index

For Product Safety Concerns and Information please contact our
EU representative GPSR@taylorandfrancis.com Taylor & Francis
Verlag GmbH, Kaufingerstraße 24, 80331 München, Germany